[redacted]

Perspectives on Genetic Discrimination

Over the past 15 years, a series of empirical studies in different countries have shown that our increasing genetic knowledge leads to new forms of exclusion, disadvantaging and stigmatization. The spectrum of this "genetic discrimination" ranges from disadvantages at work, via problems with insurance policies, to difficulties with adoption agencies.

These empirical studies on the problem of genetic discrimination have not gone unnoticed. Since the beginning of the 1990s, a series of legislative initiatives and statements, both on the national level and on the part of international and supranational organizations and commissions, have been put forward as ways of protecting people from genetic discrimination.

This is the first book to critically evaluate the empirical evidence and the theoretical usefulness of the concept of "genetic discrimination." It discusses the advantages and limitations of adopting the concept, and offers a more complex account distinguishing between several dimensions and forms of genetic discrimination.

Thomas Lemke is Heisenberg Professor of Sociology with focus on Biotechnologies, Nature and Society at the Faculty of Social Sciences of the Goethe-University Frankfurt/Main in Germany. His research interests include social and political theory, biopolitics, and social studies of genetic and reproductive technologies.

Routledge Advances in Sociology

For a full list of titles in this series, please visit www.routledge.com

84 Innovation in Socio-Cultural Context
Edited by Frane Adam and Hans Westlund

85 Youth, Arts and Education
Reassembling Subjectivity through Affect
Anna Hickey-Moody

86 The Capitalist Personality
Face-to-Face Sociality and Economic Change in the Post-Communist World
Christopher S. Swader

87 The Culture of Enterprise in Neoliberalism
Specters of Entrepreneurship
Tomas Marttila

88 Islamophobia in the West
Measuring and Explaining Individual Attitudes
Marc Helbling

89 The Challenges of Being a Rural Gay Man
Coping with Stigma
Deborah Bray Preston and Anthony R. D'Augelli

90 Global Justice Activism and Policy Reform in Europe
Understanding When Change Happens
Edited by Peter Utting, Mario Pianta and Anne Ellersiek

91 Sociology of the Visual Sphere
Edited by Regev Nathansohn and Dennis Zuev

92 Solidarity in Individualized Societies
Recognition, Justice and Good Judgement
Søren Juul

93 Heritage in the Digital Era
Cinematic Tourism and the Activist Cause
Rodanthi Tzanelli

94 Generation, Discourse, and Social Change
Karen R. Foster

95 Sustainable Practices
Social Theory and Climate Change
Elizabeth Shove and Nicola Spurling

96 The Transformative Capacity of New Technologies
A Theory of Sociotechnical Change
Ulrich Dolata

97 Consuming Families
Buying, Making, Producing Family Life in the 21st Century
Jo Lindsay and JaneMaree Maher

98 Migrant Marginality
A Transnational Perspective
Edited by Philip Kretsedemas, Glenn Jacobs and Jorge Capetillo-Ponce

99 Changing Gay Male Identities
Andrew Cooper

100 Perspectives on Genetic Discrimination
Thomas Lemke

Perspectives on Genetic Discrimination

Thomas Lemke

Routledge
Taylor & Francis Group
NEW YORK LONDON

First published 2013
by Routledge
711 Third Avenue, New York, NY 10017

Simultaneously published in the UK
by Routledge
2 Park Square, Milton Park, Abingdon, Oxon OX14 4RN

Routledge is an imprint of the Taylor & Francis Group, an informa business

Library of Congress Cataloging-in-Publication Data

Lemke, Thomas.
[Polizei der gene. English]
Perspectives on genetic discrimination / by Thomas Lemke.
pages cm — (Routledge advances in sociology ; 100)
Includes bibliographical references and index.
1. Genetic screening—Social aspects. 2. Discrimination. 3. Genetic screening—Moral and ethical aspects. 4. Human genetics. I. Title.
RB155.65.L4613 2013
616'.042—dc23
2012042912

ISBN13: 978-0-415-87858-6 (hbk)
ISBN13: 978-0-203-52611-8 (ebk)

Typeset in Sabon
by IBT Global.

Originally published as *Die Polizei der Gene. Formen und Felder genetischer Diskriminierung*, by Campus Verlag (2006).

SUSTAINABLE FORESTRY INITIATIVE Certified Sourcing
www.sfiprogram.org
SFI-01234
SFI label applies to the text stock

Printed and bound in the United States of America
by IBT Global.

Contents

Introduction

In his book *The Normal and the Pathological*, Georges Canguilhem, the French historian of science, describes the emergence of a new concept in pathology in the early 20th century: the concept of error. This concept signified a disorder which was not caused by an external pathogen, but by the body's genetic constitution. When, in 1909, Sir Archibald Garrod coined the term "inborn errors of metabolism" for inherited metabolic disorders, this proposed definition was still based on the expressiveness of a metaphor. However, in the course of the reformulation of fundamental biochemical and genetic problems in cybernetics, information, and communication theory terms during the second half of the 20th century, the idea of hereditable errors was generalized and expanded. The concept of disease has changed in the wake of the new conception of living matter as "code" or "information":

> [T]he negative of order is inversion, the negative of sequence is confusion, and the substitution of one arrangement for another is error. Health is genetic and enzymatic correction. To be sick is to have been made false, to be false, not in the sense of a false bank note or a false friend, but in the sense of a "false fold" [i.e., wrinkle: *faux pli*] or a false rhyme. (Canguilhem 1978: 172)

The introduction of error into pathology resulted both in the radicalization and the trivialization of disease. The former is due to the fact that genetic diseases affect the "organization" or "building plan" of life itself. It is not an error caused by a temporary condition, but is inherent in the subject's bodily substance. Disease, as it were, is intrinsically "inscribed" in the body and cannot be separated from it. This radicalization goes hand in hand with a de-mystification of disease, because it is conceived neither as an evil nor as an excess, but as an "original flaw in macromolecular form" (ibid.: 173). This flaw, rather than pointing to a fault or negligence on the part of the person suffering from the disease, is related to the caprices of evolution, the destiny of heredity, and the accidents of genetic read and copy errors. Canguilhem explains that, for the diseased subject, the idea of an innate organic error is anything but reassuring, because it is confronted with a disease which is experienced as "senseless," as a "fault of conduct without the fault of the conductor" (ibid.: 174). What disturbs the diseased individual beyond his or her personal fears, however, are the possible institutional responses to these "genetic errors":

> Why not dream of hunting heterodox genes, of a genetic inquisition? . . . We know that these are not only dreams for some biologists of very different philosophical persuasions, if we may call it that. But in dreaming these dreams, we enter another world, bordering on the bravest of Aldous Huxley's, from which sick individuals, their particular diseases, and their doctors have been eliminated. The life of a natural population is portrayed as a lotto sack and the functionaries delegated by the life sciences have the task of verifying the regularity of its numbers before the players are allowed to draw them from the sack to fill their cards. At the beginning of this dream we have the generous ambition to spare innocent and impotent living beings the atrocious burden of producing errors of life. At the end there are the gene police, clad in the geneticists' science. (Ibid.: 174–175)[1]

One may wonder whether today we have reached the age of the "gene police" predicted by Canguilhem in an almost visionary fashion.[2] A series of empirical studies conducted in different countries in the last twenty years has pointed to new forms of exclusion, disadvantage, and stigmatization caused by the increase of genetic knowledge (Billings et al. 1992; Geller et al. 1996; Low, Kind, and Wilkie 1998; Otlowski et al. 2007; Taylor et al. 2008). The spectrum of this "genetic discrimination" ranges from labor market exclusion and problems with insurance policies to difficulties involving adoption agencies. Thus, applicants' job qualifications have been rejected, and they have been denied employment on the grounds of a possible future disease. Similarly, health and life insurance companies have cancelled or refused to issue policies when genetic risks were suspected among their (potential) customers. In other cases, couples were denied child adoption when one spouse had a disposition for a genetic disease. Experiences of genetic discrimination have also been documented with regard to the health sector, the education system, and the military.

The hopes of a "molecular medicine," able to identify and prevent disease before concrete symptoms manifest themselves, are overshadowed by the nightmare of a society in which career prospects, insurance options, educational pathways, and family planning are determined by genetic characteristics. The academic discourse surrounding the negative consequences of growing genetic knowledge is dominated by the fear of a "genetic underclass" (Nelkin and Tancredi 1994: 176; Nelkin 1992: 190; Keays 2000: 84–85; Kutter and Ruess 2001) or "new pariahs" (Kitcher 1996: 127)—individuals stigmatized, pathologized, and discriminated against on the basis of their genetic properties. The main concern is that innovations relating to genetic knowledge will produce new types of class formation not based on socio-economic status, but on genetic criteria. A danger is seen in a "genetocracy with individuals, ethnic groups, and races increasingly categorized and stereotyped by their genotype, making way for the emergence of an informal biological caste system in countries around the world" (Rifkin 1998: 3; cf. also Dickens 2001: 105–106).

Apart from academic discourse, the problem of genetic discrimination has now prompted regulatory initiatives at the juridical and political level to determine the conditions and limitations concerning the use of genetic information. Many countries have introduced laws designed to prevent the discriminatory treatment of individuals due to their genetic properties. These legislation projects guarantee the right to self-determination and "genetic privacy," and protect personal data against misuse. Moreover, the issue of genetic discrimination is being taken up in the media and the arts. It is addressed in newspapers and magazines, and in television and radio reports. Science fiction movies like *Gattaca* and novels such as *Beggars in Spain* describe the repercussions genetic segregation and selection have on societies in which key social positions are exclusively taken by the "genetically enhanced."[3]

Although genetic discrimination appears to be a straightforward and simple issue, it is in fact a rather complex and under-specified phenomenon. Still largely unknown only a few years ago, the notion has now become an ambivalent term used in many different contexts and for diverse purposes.[4] Many scholars have remarked critically that the term "genetic discrimination" is not employed with sufficient precision and that its conceptual and normative premises are rarely made explicit.[5] Accordingly, the spectrum of proposed definitions and patterns of reasoning is broad. It ranges "from the allegation of irrationality and lack of objectivity in the assessment of 'genetically handicapped' individuals to the proposition of the invasion of morally justified personal privacy to the assertion of eugenic intentions" (Paslack and Simon 2005: 133–134).

The fact that this "new form of social prejudice" (Rifkin 2000) is an empirically diverse and complex phenomenon can be illustrated with the help of three exemplary cases in which the accusation of "genetic discrimination" was made.

The first one occurred in Denmark about ten years ago. In his production of "Hamlet" at the Royal Danish Theater in Copenhagen, the director wanted to cast an actress living with Down's syndrome as Ophelia. Because some members of the ensemble were of the opinion that it was dramatically incoherent as well as aesthetically inappropriate to cast a disabled woman in the role, the production had to be cancelled at the last minute. This sparked off a fierce debate in the Danish media about the pros and cons of this decision, in which the famous film director Lars von Trier also took part. Von Trier accused those actors refusing to accept the director's original decision of "genetic discrimination," because, he argued, they wanted to exclude an actress from the role and based this decision upon her genetic characteristics (Tolmein 2002).

The protagonist of the second case is the popular American basketball player for the Chicago Bulls, Eddy Curry. During a medical examination, Curry was diagnosed with an irregular heartbeat and an enlarged heart. The doctors found that both these factors indicated a heightened risk of hypertrophic cardiomyopathy (HCM). HCM is a cardiac problem that is believed to have been the cause of death of two basketball players who collapsed and died on the court. Following the diagnosis, the Chicago Bulls management suspended Curry from play and requested him to submit to a series of tests, including a DNA test. Although these tests cannot verify the disease, they do allow the identification of a possible risk of being affected by the disease. The player, however, refused to submit to this request. Consequently the management announced they would not consider him for further games until he agreed to the tests (Bodie 2005).

The third case took place in Germany. In August 2003, a teacher was denied employment as a civil servant on probation in the German *Land* of Hesse. In response to a physician's questions, the young woman declared that her father had Huntington's disease (Morbus Huntington)—a neurogenerative disease that manifests itself late in life. The medical report came to the conclusion that at present the applicant was healthy; her civil servant status, however, was denied with the justification that a high probability existed that she would be affected by the disease in the future and thus become permanently unfit for work (Traufetter 2003).[6]

The teacher from Hesse, the basketball player from Chicago, and the Danish actress living with Down's syndrome—despite the differences in the cases, the common denominator of these experiences is that they are described, by the people affected themselves or by others, as "genetic discrimination." Of course, this classification does not *per se* imply morally reprehensible behavior. "Genetic discrimination" is instead a normatively very ambivalent phenomenon open to a multitude of diverse, indeed even contradictory assessments and interpretations. In each of the three examples, the direction of moral judgment, at first seemingly indisputable, can just as well be reversed. Thus one could ask, with regard to the first case, whether genetic discrimination had also, if not especially, been involved, that is, whether the actress had been cast for the character solely because of her disability (Tolmein 2002). Looking at the second case, it may conceivably be regarded as thoughtful and appropriate for the employer to show concern for his employee's health and to prohibit any life-endangering practices, at least within the context of his sports team—not least to circumvent future damage claims. It may also be deemed proper and advisable to subject—as in the third example—civil servants to rigorous medical tests in order to bar individuals with increased health risks from the civil service, due to the special care and loyalty relationship that it represents in Germany.

The present book addresses the complexity and the manifold dimensions of the problem of genetic discrimination. I will touch upon normative issues occasionally, but my primary concern is the unveiling of terminological and theoretical shortcomings and empirical gaps that characterize the current discussion. It is amazing that, despite the ubiquitous calls for regulations and bans on misuse, the heated media discourse, and the differentiated and elaborate nature of bioethical considerations and deliberations, there is so little empirical knowledge about the forms and fields of genetic discrimination. Beyond cultural representations, media debates, political regulations, and academic controversies, fundamental questions remain unanswered: What is genetic discrimination? How is it different from other forms of disadvantage, disregard, and stigmatization? How widespread are practices of genetic discrimination? Who is discriminated against, in what way, and based on which genetic properties?

This study summarizes the results of the international academic and legal-political discussion of the phenomenon of genetic discrimination up to now, identifying its strengths and weaknesses and suggesting a more extensive and differentiated concept of discrimination. So far, the theoretical discourse on the phenomenon has been dominated by scholars of the natural sciences, medicine, and legal sciences; this study has an explicitly sociological angle of analysis. It seeks to investigate "genetic discrimination" within a framework that also integrates the social and institutional constituents and consequences of the "molecularization of medicine." In this way, the book contributes to the sociological analysis and critique of genetic discrimination and establishes a framework for further empirical investigations and theoretical reflections of the issue. In order to investigate the problem of genetic discrimination, I will make use of an analytics of government inspired by the work of the French historian and philosopher Michel Foucault (Foucault 2007, 2008; Lemke 2004b; Rose 2007; Rouvroy 2008; Bröckling, Krasmann, and Lemke 2011; Lemke 2011).

The first chapter examines how the vision of "molecular medicine" was able to take shape in recent decades. Whereas genetic aspects only played a subordinate role in medicine until well into the second half of the 20th century, genetics is now ascribed a pivotal role in the diagnosis, prevention, and treatment of many common diseases. The chapter pursues the thesis that *one* of the central prerequisites for this paradigm shift was the conceptual expansion and redefinition of "genetic disease." Today, this concept embraces not only monogenetic diseases, but increasingly also multifactorial illnesses such as cancer, Alzheimer's disease, diabetes, and so on. Moreover, the idea of a molecular medicine is built upon a redefinition of disease as a genetic anomaly. As this happens, genetic predispositions and risks which have not (yet) clinically manifested themselves come into medical view. As a consequence, people can be regarded as "ill" or "disabled" and discriminated against based upon their genetic properties.

The second chapter starts by presenting a summary of the most important empirical studies on experiences of genetic discrimination. In addition to providing an overview of the state of international research, this outline illustrates the breadth and diversity of discriminatory practices. For the most part, the investigations are from the United States. However, studies from the United Kingdom and Australia as well as reports on individual cases from other countries will also be presented. The second part of this chapter is devoted to the legal regulation of genetic discrimination. A brief introduction to legislative initiatives and positions on the part of international and supranational organizations and commissions will be followed by a critical appraisal of the German Genetic Diagnostics Act (Gesetz über genetische Untersuchungen bei Menschen, GenDG), which came into force on February 1, 2010, and the Genetic Information Non-Discrimination Act (GINA) in the United States. Finally, data protection issues and practical problems involved in the legal fight against genetic discrimination will be discussed.

The third chapter demonstrates the theoretical, normative, and practical imprecision and contradictions of the concept of genetic discrimination as it is used in scholarly studies and political and legal discussions. These range from discrepancies in the use of the concept of discrimination in research practice, via empirical deficits that disregard key arenas of genetic discrimination, to conceptual problems in arguing for an exceptional status for genetic information. The chapter shows that the idea of a specific protection of genetic data as opposed to non-genetic information fosters wrongful and unequal treatment of the individuals involved. Genetic discrimination is generally regarded as more problematic in ethical and legal terms than the inequitable treatment of the sick and disabled. The reason for this lies in a genetic essentialism that ascribes an overstated importance and autonomy to genetic factors. Paradoxically, the theoretical analysis as well as the political and legal criticism of practices of genetic discrimination often rests upon the implicit notion that genes determine human existence in a fundamental way and that they constitute the nucleus of the personality. The underlying basis of the discourse is the spurious concept of a genetic program which is supposedly responsible for individual development as well as personal identity.

In contrast to the United States, the United Kingdom, and Australia, no empirical studies on experiences of genetic discrimination have yet been conducted in Germany. Chapter 4 will present the findings of the first explorative study undertaken by the author in 2004. Within the context of this study, individuals living with Huntington's disease were asked about their experiences of disadvantage, stigmatization, and disregard. Based upon case examples, the study demonstrates that the concept of genetic discrimination underlying academic discourse is too narrow. It is reserved for types of discrimination and unequal treatment in relation to the genotype, yet excludes to what extent those already living with the disease are discriminated against based upon the genetic nature of their illness. What is more, the accounts illustrate that beyond practices of organizational discrimination—the focus of studies conducted in this field up until now—the forms of interactional and indirect discrimination need to be more systematically integrated into the analysis.

Chapter 5 analyzes the focus on genetic factors in medicine as a "truth regime" representing a particular configuration of medical knowledge, enabling the development of new forms of intervention in relation to the prevention, diagnosis, and treatment of diseases. Within the current social and political constellation, genetic medicine not only reinforces the tendency for causes of diseases to be sited in the individual body; in addition, the danger looms ahead that responsibility and costs for diseases will be increasingly assigned to the person directly affected. The search for the genetic causes of diseases does not lead to a negation or limitation of the individual's accountability; rather, the opposite is the case. I argue that the recently acquired genetic knowledge enables the extension of existing obligations and the establishment of new areas of responsibility. The expansion of "genetic responsibility" will be more closely investigated in two fields. The idea of "informational responsibility" toward relatives, that is, the expectation that they should be informed about potential genetic risks, contradicts the protection of privacy and the right to self-determination. The demand for "self-responsibility" in dealing with genetic risks undermines the right not to know. As shown by the review of legal cases in the United States in recent years, there is an observable trend towards the institutionalization of genetic responsibility, both with regard to obligations to disseminate genetic knowledge to relatives and to seek information on one's own genetic risks.

Chapter 6 will summarize the conceptual, empirical, and theoretical findings of this study. It also points out the paradoxes of some critical accounts of genetic discrimination. The key insight is that the analytical and political concentration on practices of genetic discrimination promotes the very "geneticization" (Lippman 1991) of the body, disease, and deviance which once was the motive for fighting genetic discrimination. Criticism directed at genetic discrimination commonly starts with the assumption that genetic factors play an exceptional role and rests upon the fiction of a "genetic norm," which is made to appear an immanently given and neutral biological fact. As a consequence, the systematic connections between already existing forms of exclusion, stigmatization, and discrimination, on the one hand, and genetic modes of reasoning, on the other, are neglected, while simultaneously the erroneous assumption of the genetically normal is confirmed.

The present text is an updated and expanded version of a book published in 2006 in German by Campus under the title *Die Polizei der Gene. Formen und Felder genetischer Diskriminierung*. It was based upon articles published in various journals and books over a number of years. A modified version of the first chapter was published in *Prokla. Zeitschrift für kritische Sozialwissenschaft* (volume 33(3), 2003, pp. 471–492) under the title "Molekulare Medizin? Anmerkungen zur Ausweitung und Redefinition des Konzepts der genetischen Krankheit." Parts of the second chapter feature in the article "Formen und Felder genetischer Diskriminierung: Ein Überblick über empirische Studien und aktuelle Fälle," co-authored with Christiane Lohkamp (in: Wolfgang van den Daele [ed.], *Biopolitik*, Wiesbaden: Verlag für Sozialwissenschaften 2005, pp. 45–70). Versions of chapters three and four were published in the journal *Soziale Welt* ("'Die Polizei der Gene'—Genetische Diskriminierung und die Fallstricke der Kritik," in: *Soziale Welt*, volume 56(1), 2005, pp. 53–72 and "Genetische Diskriminierung in Deutschland. Eine explorative Studie am Beispiel der Huntington-Krankheit," in: *Soziale Welt*, volume 56(4), 2005, pp. 417–440) and in *Genomics, Society and Policy* ("Beyond genetic discrimination: Problems and perspectives of a contested notion," volume 1(3), 2005, pp. 22–40 and "'A slap in the face': An exploratory study of genetic discrimination in Germany," volume 5(2), 2009, pp. 22–39). Chapter 5 is an expanded version of a contribution to an edited volume ("Lebenspolitik und Biomoral: Dimensionen genetischer Verantwortung," in: Heinrich-Böll-Stiftung [ed.], *Die Verfasstheit der Wissensgesellschaft*, Münster: Verlag Westfälisches Dampfboot 2006, pp. 332–345).

The study is based upon investigations conducted within a research project funded by the German Research Foundation under the title "Genetische Diagnostik in der Risikogesellschaft" and carried out at the Institute for Social Research in Frankfurt am Main. Many of the ideas and inspirations that made this book possible originated in discussions with members of the Genetic Screening Study Group in Boston and with representatives of the Council for Responsible Genetics in Cambridge (Massachusetts). I would like to express my gratitude to Joseph Alper, Jon Beckwith, Peter Conrad, Lisa N. Geller, and Sujatha Byravan, as well as to Diane Paul and Sarah Jensen. For helpful comments and valuable inspiration, my thanks go to my colleagues at the Institute of Social Research, as well as to Anne Waldschmidt, Kathrin Grüber, Susanne Krasmann, Sigrid Graumann, Ulrich Bröckling, Annette Prassel, Barbara Prainsack, Ine van Hoyweghen and two anonymous reviewers of the manuscript.

I would like to thank David Sánchez and Anja Löbert for providing a first translation of the original version of the text. This book received a grant from the Volkswagen Foundation, and I am deeply indebted to the organization for making this publication possible. Jonas Rüppel, Shirin Moghadda and Corina Färber were of enormous help in revising and updating the manuscript, Katharina Hoppe was responsible for preparing the index and Gerard Holden attentively read and corrected the manuscript. Finally, I am particularly grateful to all those who have shared their experiences of genetic discrimination with me, even though this was often painful for them—for reasons that will be examined in this study.

1 Disease as Error?
Foundations and Limits of Molecular Medicine

> In the future health will be fundamentally linked to knowledge about genetically conditioned predispositions, at the individual as well as at the social level. It is foreseeable that this genetic lifeworld will become just as *true* for people as the hygienic-bacteriological lifeworld has become for us. (Labisch 2002: 135–136; emphasis in original)

In the history of medicine there have always been etiological concepts that have assumed a paradigmatic function in research and clinical practice for the entire field of medicine or in key subdisciplines. Whereas medieval medicine followed Galen's humoral pathology to a large extent, after the discoveries of Pasteur and Virchow in the 19th century medical interest concentrated on fighting infectious diseases. At present a new medical paradigm is emerging, which ascribes decisive significance to genetic factors in pathogenesis and conceives of disease as an error or mistake in the "building plan of life" (Canguilhem 1978: 171–174; Foucault 2003; Weigel 2002).[1]

As popular and plausible as the idea of a "genetic medicine" appears today, it is in fact relatively recent. In the first half of the 20th century, medical aspects still played a minor role in genetics. The field was primarily of interest to zoologists and botanists, as the most impressive medical successes of this period came above all from improvements in labor, living and nutritional conditions, as well as from the development of vaccines and antibiotics (Rushton 1994; Porter 1997). Even after the introduction of genetics in medical faculties in the second and third quarters of the previous century, its clinical relevance remained limited to comparatively rare so-called hereditary diseases. The situation changed dramatically in the wake of the Human Genome Project, as many scientists proclaimed an "era of molecular medicine" (Ganten and Ruckpaul 2001: 3; cf. Caskey 1995; Clark 1997; Williams and Hayward 2001). Increasingly, genetic explanations and concepts were assigned central significance in the diagnosis, prevention, and therapy of many common diseases. Today, genetic testing and screening are available not just for a few hereditary conditions, but also for diseases with a high prevalence. As the scientific and medical focus shifted to the presumed genetic "factors" of diseases, a new orientation in public health policy and education emerged. This would use genetic information in population screening to determine individual susceptibilities to common disorders like cancer, heart disease, or diabetes, and to identify groups at risk, so that preventive interventions could be initiated (Khoury, Burke, and Thompson 2000, Brand, Brand, and Schulte in den Baumen 2008). This proposal is not limited to representatives from human genetics or medical fields, but is also shared by politicians. Presenting the draft version of the human genome on June 26, 2000, President Bill Clinton declared in a joint press conference with Prime Minister Tony Blair: "Today, we are learning the language in which God created life . . . With this profound new knowledge, humankind is on the verge of gaining immense new power to heal." (*New York Times* 2000) In a similar vein, the former German Federal Minister for Education and Research, Edelgard Bulmahn, was convinced that "the application of modern molecular-biological methods and insights will revolutionize daily medical practice" (Bulmahn 2003: 6).

How was it possible for this profound change to take place in the last thirty years? What factors promoted the "geneticization of medicine" (Lippman 1991) in this comparatively short period of time? This chapter will propose an answer to this question, suggesting that *one* central condition for this far-reaching paradigm shift lies in the expansion and redefinition of the concept of genetic disease. The following argument will therefore focus on the analysis of the historical and epistemological conditions for the current orientation of medicine and the increasing significance of genetic knowledge as a way of determining health and illness.

SCIENTIFIC ACHIEVEMENTS, TECHNICAL ADVANCES, CONCEPTUAL INNOVATIONS: THE ESTABLISHMENT OF THE "GENETIC POINT OF VIEW"

A series of historical conditions of emergence accounts for the growing impact of genetic knowledge in medicine. One seemingly paradoxical reason for this phenomenon lies in the success of the paradigm that preceded the genetic discourse: the bacteriological paradigm. The discoveries of Louis Pasteur and Robert Koch not only signaled the beginning of laboratory medicine, but also helped to strengthen the scientific basis of the concept of disease (Latour 1988). Bacteriological knowledge entered medical practice, resulting in an overall decrease in the mortality and morbidity of infectious diseases and an increase in general life expectancy. Consequently, medical interest shifted to chronic and widespread conditions such as cancer, diabetes, and cardiovascular diseases (Koch 1993: 5–7; Labisch 2002).

Thanks to the remarkable technical and scientific advances made between the 1950s and the 1980s, genetics moved from a comparatively insignificant discipline to the center of the biosciences in the second half of the 20th century (Kay 2000).[2] Not only did its relative significance within biology increase during this shift, but genetics also went through an inner transformation which ultimately opened it up for medical problems. The development of genetic engineering in the 1970s transformed the explorative discipline, until then dedicated to basic research, into an applied science with potentially important medical implications (Lindee 2005).

The historian of science Hans-Jörg Rheinberger has described this shift in detail. He points out that the traditional biophysical, biochemical, and genetic techniques used in molecular biology until the 1960s aimed for the construction of an experimental environment, which allowed cellular structures to be recognized and analyzed via work on "model organisms" (Rheinberger 2000). According to Rheinberger, the manner in which the molecular structures and processes of the organism have been opened up for intervention has been fundamentally transformed since the 1970s. The introduction of recombinant DNA techniques was a decisive factor, because their key tools, such as restricting, transcribing, and replicating enzymes, do not depend on analytic or electronic machines but are themselves biological macromolecules. To the extent that the DNA of an organism can be modified with the aid of its own (un-)changed components, molecular biology, in the form of molecular engineering, leaves behind the traditional paradigm and establishes a new relation between molecular biology and medicine:

> The intact organism itself is turned into a laboratory. It is no longer the extracellular project, i.e., the deliberative 'rewriting' of life. From an epistemic perspective, this procedure makes the practice of molecular biology, *qua* molecular engineering, substantially different from traditional intervention in the life sciences and in medicine. This intervention aims at *re-programming* metabolic actions, not just interfering with them. (Ibid.: 25; emphasis in original; cf. Rheinberger 1997b)

However, the significance of genetics for medicine did not only grow due to technical innovations and scientific achievements; at the same time, the comparatively rigid concept of genetic causation was subjected to a profound revision. In the light of techno-scientific advances, the relationship between genetic and environmental factors has been redefined in molecular biology since the end of the 1960s. Genes increasingly appeared open to technical interventions and scientific innovations. The focus on genes or molecular structures thus no longer implied fatefulness and unchangeability; on the contrary, genetics presented itself as a privileged field for techno-scientific interventions:

> Where the traditional view had been that 'nature' spelled destiny and 'nurture' freedom, now the roles appeared to be reversed. The technological innovations of molecular biology invited a vastly extended prowess, encouraging the notion that we could more readily control the former than the latter . . . [I]n the vision inspired by the successes of molecular biology, "nature" became newly malleable, perhaps infinitely

> so; certainly it was vastly more malleable than anyone had ever imagined "nurture" to be. (Keller 1992: 288–289.; cf. Paul 1998a)[3]

In spite of these scientific and technological advances, genetic issues only played a marginal role in medicine up to the 1970s, as the president of the *American Society of Human Genetics*, Barton Childs, soberly noted in his speech at the society's annual assembly in 1976.[4] In the text of his speech, Childs lamented the fact that genetic knowledge generally received little attention from the medical profession and medical education, and that the institutional and conceptual reorganization of medicine, necessary in his opinion, had yet to be accomplished. He referred to a survey among practicing physicians on genetic screening, which demonstrated that in the mid-1970s, less than half of the physicians surveyed regarded genetic diseases as a serious issue for their clinical work (Rosenstock et al. 1975). According to Childs, four reasons above all explained why physicians did not (yet) think in genetic terms. First, the work of doctors was traditionally focused more on individuals than on families. Second, medical practice had been more concerned with individual cases of disease than with general prevention and public health education. Third, genetic counseling depended heavily on research and was only carried out at specialized genetic clinics. Finally, doctors concentrated on the acute treatment of patients, while broader public health issues were largely excluded from clinical work. For these reasons, a person with a genetic disease was "seen to be the concern of the practitioner, but the dissemination of a mutagen which may lead to new examplae of that same disease is seen as the business of some public health agency, while the spread of a mutant through a population is generally perceived to be no one's affair" (Childs 1977: 7; cf. Yoxen 1984: 50–52; Koch 1993). The "genetic point of view" called for by Childs (ibid.: 8), which should be anchored and cultivated among practicing physicians and medical students, would allow them to provide a better explanation for the differences and similarities among persons suffering from the same disease and extend the medical gaze beyond the individual to his or her family and even to the whole of humanity (ibid.: 8–9).

The extremely limited reception of genetic modes of reasoning and of genetic technology in the medical practice of the 1970s contrasts sharply with today's "era of molecular medicine." In light of the scientific achievements of genetic research, representatives of human genetics often request that they should be accorded a central position within medicine. Some years ago the German human geneticist Holger Höhn pointed out in an article titled "The Genetification of Medicine and the Future of Human Genetics" that an increasing number of clinical disciplines are integrating genetic methods and models into their work. The "exotic subject" has become a "flourishing science," indispensable for specialized fields of medicine: "If we use the opportunity correctly then the 'geneticist' in the gynecological clinic, in neurology, psychiatry, pathology, pediatrics, ophthalmology, etc. will be someone who has gone through our school . . . The motto must be 'our woman, our man in the clinic'" (Höhn 1997: 174; cf. also Schmidtke 1998; Vogel 1998). Since the 1970s, the "genetic point of view" has become established in such a profound manner in medicine that a quarter of a century after his 1976 speech, Childs announced the synthesis of genetics and medicine. This, he argued, was giving rise to a new "logic of disease" (Childs 1999) that engendered enormous diagnostic potential, as well as preventive and therapeutic opportunities: "Spontaneously and gradually, a synthesis of genetic and medical ideas . . . is now beginning. It is a wedding of equals, and its offspring affords us a new logic, a new way of examining old questions of the nature of disease and its causes." (Ibid.: x)

In the past thirty years, genetics has succeeded in establishing itself as the fundamental science of medicine, which should not just be authoritative in relation to rare diseases but also with regard to the development and manifestation of diseases in general. The idea that "all human disease is a genetic phenomenon subject to technological control" (Lindee 2005: 2) has gained considerable public legitimacy and scientific currency in the wake of the Human Genome Project. It is claimed that "genetic medicine" opens up a more exact analysis of pathogenesis via a new taxonomy of diseases, because conditions only phenotypically describable until then could be explained and understood through genotypical modifications (Raem et al. 2001; Ganten and Ruckpaul 2001: 8; cf. Kaplan and Junien 2000). The goal of this new orientation within medicine is to establish a model of verification that no longer determines diseases according to their clinical symptoms, but instead traces them back to their genetic "causes." This new approach entails the hope that the classification according to genetic criteria will make it possible to explain the multifarious variations of these diseases, just as much as the apparent differences in their symptoms, in their clinical course and in their reactions to pharmaceuticals and forms of therapy. In this perspective, molecular genetics is to provide the tools to define diseases according to the mechanisms they are based on, leading "to an entirely new understanding of medicine in many areas, which will also be relevant for the clinical doctor and the patient in the foreseeable future. The analysis of disease symptoms will be increasingly augmented by a genetic-causal aspect." (Maurer and Lehrach 2001: xiii; cf. also Bell 1998: 618; critically: Marteau and Holtzman 2000)[5]

The expansion of the concept of genetic disease and its redefinition were decisive for the growing relevance of molecular-biological knowledge to medical research and clinical practice. This process allowed the traditional medical focus to be kept on the individual and at the same time—as already called for by Childs in the mid-1970s—made it possible to include in it considerations on public health policies and options for disease prevention.

SYMPTOMATOLOGY AND VERIFICATION

Since the beginning of the 1990s, genetic factors have been determined as being at the root of an increasing number of diseases. This conception of disease is based on the assumption that one or more "defect" genes lead to the loss or hindrance of a metabolic function, which ultimately is causally responsible for the disease or makes the organism more susceptible to certain diseases. The evolution toward a "geneticization of medicine" (Lippman 1991) can be illustrated by two key indicators. For fifty years, diseases presumed to be hereditary have been registered in the so-called McKusick Catalogue. The *Mendelian Inheritance in Man* (MIM) is today the most important medical database providing online information on human genes and the number of genetic disorders. It is permanently kept up to date. The number of registered clinical pictures for which genetic causes are known amounted to 5,000 in 1992, rose to 10,000 at the end of 1998, and today lies at nearly 22,000 entries (May 4, 2012: 21,199; cf. http://www.omim.org).[6]

Just as impressive as the rapid increase in the number of genetic diseases is the growth of the concept of genetic risk in medicine. The database PubMed, in which all the leading medical journals are indexed, documents an astonishing rise in the number of articles dealing with genetic risks in the last forty years.[7] Whereas in the period between 1967 and 1976 only thirty-five articles were listed that contain the terms "genetic risk" in their title or abstract, ten years later, the figure was already 160 (1977–1986); between 1987 and 1996, this rose to 644, increasing once more in the following decade to 2,913 (1997–2006).[8] In other words: in the last forty years, the number of articles in medical journals concentrating on the analysis of genetic risk has multiplied by a factor of more than eighty.[9]

The indicators mentioned illustrate how within medical research more and more causes of disease are being traced back to genetic mechanisms. Two trends can be observed here. First, we can discern an extension of the concept of genetic disease. Genetic factors are not merely responsible for "monogenic" diseases but also for multi-factorial illnesses such as cancer, Alzheimer's, and diabetes.[10] Second, there is a qualitative shift in and redefinition of the concept of genetic disease. Nearly thirty years ago, the historian of science Edward Yoxen pointed out that the relationship between human genetics and medical science had changed profoundly. In this process of redefinition, diseases were often understood as deviations from the genetic norm, as a result of which the social and medical significance of molecular genetics had grown considerably (Yoxen 1984; cf. Koch 1993). In the following years, "genetic disease" became an extremely large category (Keller 1992: 292), confusing the various causal relationships and conceptions of disease with one another: "(1) abnormalities having a genetic cause, but not leading to any physiological or behavioral 'disorder', which we may call genetic disorders; and (2) genetic disorders that are apprehended as diseases and (3) disorders that are neither genetic nor diseases, even though they may be labeled as such by some people." (Yoxen 1984: 49) In this manner, a productive circle has been established that infers diseases from genetic abnormalities and traces diseases back to genetic abnormalities.

The redefinition of diseases as genetic disorders does not just lead to an expansion in the number of genetic diseases. It also raises the question of the meaning and limits of genetic factors in disease causation. When potentially every disease can be traced back to a genetic modification, one might ask what genetic factors exactly contribute and how genetic causes of disease are to be distinguished from non-genetic ones.[11] The growing establishment of genetic medicine is linked to two significant semantic shifts (cf. also Paul 1999: 160–161). First, the category of "genetically determined" has expanded so considerably that it does not merely include behavioral disorders and forms of disease passed on through the germ line, that is, "genetically determined" factors in the traditional sense of the word (Keller 2000: 158, FN. 37). The concept of genetic disease today also includes so-called somatic mutations, that is, mutations that appear spontaneously and are acquired during the course of life. From this perspective it can be stated that "non-hereditary diseases can also be genetically determined" (Schmidtke 1998: 12). Thus, every type of cancer for example can be regarded as a genetic disorder, because cancer is always associated with a collapse of the "normal" genetic regulation.[12]

Second, the "genetic point of view" allows the concept of disease to be extended to conditions and variations of the norm that until recently were not regarded as pathological. When genetic modifications provide the basis for determining diseases, then genetic dispositions, risks, and so on which have not (yet) clinically manifested themselves will also come under the medical gaze. The model of a predictive medicine is thus outlined, one which seeks to uncouple itself from the observable or empirically determinable symptoms of a disease. At the level of the genotype, a disease does not necessarily have to manifest itself in a clinical picture; on the contrary, the goal of molecular medicine is to prevent the emergence of specific symptoms (Koch 1999; Kollek and Lemke 2008). Furthermore, the "genetic point of view" relies on the idea of two fundamental dimensions of the body and systematically separates phenotype and genotype. In this light, the phenotype is understand as a more or less "passive" recipient of more profound activities and processes on the genetic level, and is conceived of as the mere "expression" or "effect" of the genotype. This conception grants genes a specific persistence and stability quite distinct from the suffering and perishable body. They are no longer considered as integral elements of the body; rather, the living body is perceived as the vehicle of the genes. As a result, the concern for possible health risks (and the need for surveillance and control) not only affects the individual body but also extends to family members and future generations (Gudding 1996; Hallowell 2000; Kollek 2002: 111–115).

Consequently, a double epistemological basis for the extension of the concept of "genetic disease" can be observed. The qualitative redefinition of diseases according to genetic criteria is complemented by their cognitive "virtualization," by the uncoupling of the disease from "its" symptoms. Moreover, this process of analytic separation and differentiation is even regarded as proof of a scientific approach. The underlying assumption is that the "unclear" symptoms at the phenotype level obscure the view of the "real" genetic mechanisms, and only understanding the latter permits an exact differentiation and systematization of the causes of the disease. In this manner, however, people are considered and treated as "ill" who are completely free of any symptoms and whose "disease" is only determined at the molecular level.

WHAT IS A GENETIC DISEASE?

What makes a certain anomaly or observable functional disorder a genetic disease? This question is anything but trivial, as even textbooks on medical genetics contain different, often incompatible definitions of disease.[13] From a historical perspective, it is important to keep in mind that some dysfunctions were seen as diseases before they were regarded as genetic (one example is sickle cell anemia). Conversely, certain abnormalities were classified as genetic before they were regarded as diseases (Yoxen 1984: 48–49).

In an instructive article, Kelly C. Smith discusses the question of in what sense diseases can be regarded as genetically caused or "determined" (Smith 2001). He criticizes a number of interpretative models which, for various reasons, he believes to be insufficient or misleading.[14] One prominent proposal for a definition states that a disease is always to be termed genetic when genetic factors play a causal role in its pathogenesis. As Smith demonstrates, this assumption seems plausible at first but ultimately produces no gain in knowledge, as it is true but empty of content. In this broad sense, it can be claimed of any disease that it possesses the quality of being "genetic," simply because genes control the production of the proteins essential for the life of any organism. According to such an all-encompassing definition, even lead poisoning could be understood as a "genetic disease," because people possess widely different (gene-based) mechanisms in their reactions to toxic substances.

The second approach presented by Smith does not focus on the mere contribution of genes to pathogenesis, but instead aims at the aspect of regulation and control: "The manipulability criterion says that a disease is genetic if and only if it is best controlled (prevented) through manipulation of the genes" (Ibid.: 22). Yet this attempt at a definition also raises problems. The question arises of what "best" is supposed to mean, because a specific disease can be treated in many different ways. Smith cites the example of cystic fibrosis.[15] This disease can be "healed" through respiratory exercise and medication, or in the future possibly by means of genetic intervention. What "best" means in each case, however, is not a function of technical and scientific progress but depends on social values, cultural patterns, political interests, financial restrictions, commercial incentives, and so on.

Rogeer Hubertus Maria Vincent Hoedemaekers, after an extensive discussion of the dominant concepts of disease, comes—like Smith—to the conclusion that only in very few cases is it appropriate and helpful to speak of "genetic diseases." He recommends reserving the term for "cases where it is clear that genetic factor(s) are both necessary and sufficient determinant(s) for disease" (Hoedemaekers 1998: 95), that is to say, for those rare monogenic diseases that manifest themselves in a certain set of symptoms, for the most part independent of environmental factors. In all other cases, he considers it preferable to speak of "genetic risks." Hoedemaekers would thus like to see the term genetic disease limited to those genetic anomalies which follow a strictly deterministic model in the disease expression; by contrast, the term "risk" should simply describe an increased probability of illness (ibid.: 79–99; cf. also Hoedemaekers and ten Have 1999).[16]

The typical example that is repeatedly cited for the prospects of genetic medicine is the successful treatment of phenylketonuria (PKU), a metabolic disorder in which an enzyme deficiency prevents the metabolism of amino acid phenylalanine into tyrosine. Nonetheless, a thorough analysis of the history of PKU treatment actually demonstrates the limits of an approach that concentrates on genetic mechanisms and options for intervention alone.[17] Although it is possible to describe PKU as a genetic disease, the metabolic disorder can be prevented by means of a specific diet. To say that a disease is genetically determined does not automatically mean that it is advisable or sensible to focus on genetic factors for its prevention and therapy. A pragmatic attitude is preferable, one that aims less at a complete analysis of the causes of a disease and more at the selective interruption or control of the symptoms. Strategies for disease prevention therefore do not necessarily rest on an extensive and exhaustive analysis of pathogenesis, but on a determination of the points of intervention which effectively stop the emergence of pathological symptoms. In this sense a disease can be prevented in more than one way—even when it concerns a "hereditary disease" such as PKU (cf. Willett 2002).[18]

If this assessment is correct, then the central status that is repeatedly claimed for genetics in clinical medicine has to be doubted. Even if molecular genetics were to fulfill its ambitious claims and provide significant new insights into the courses and causes of a disease, it remains to be seen if the "genetic point of view" will have the promised major significance for clinical research and the development of medical therapies. The diagnosis of a genetic disease (whatever this, in light of the preceding discussion of the different interpretations, might specifically mean) in no way determines that it must be genetically treated, because within the framework of clinical work, the focus is not on finding the "causes" of diseases but rather their "Achilles' heel" (Rees 2002; Smith 2001: 19–20).[19]

One must keep in mind then that the assignment of genetic factors as fundamental to explaining the emergence and the course of diseases does not necessarily mean they can only be genetically diagnosed or treated. Yet the definition and diagnosis of genetic diseases raises a more fundamental problem: the question of which genetic variations are regarded as pathological and which as medically irrelevant.

VARIATION AND COMPLEXITY

The goal of molecular medicine is the ability to establish a scientific (or more specifically, genetic) classification of diseases. The detailed description of the molecular causes of disease—according to the central premise—would allow medical research to develop better and more precise preventive measures and therapeutic techniques, because they are based on knowledge of the genotype.

This vision, however, confuses two analytical levels which should remain distinct. At the level of the genome, there are no diseases; only individual variations are scientifically determinable. Although genetic textbooks (e.g. Strachan and Read 1996: 453; Ramser and Meindl 2001) and scientific journals (e.g. Balding and Richardson 2000; Robinson, Krawitz, and Mundlos 2011) repeatedly refer to "disease genes," this term is in fact misleading and ultimately false, for only entire organisms can be ill. One cannot speak scientifically about "sick" genes. The characterization of certain conditions as "ill" or "abnormal" is also a social and moral definition, which signals "a need for action" and initiates as well as legitimates intervention. The identification of diseases is therefore not a purely descriptive process, but instead always implies a normative assessment on the need for a change of condition.[20] The determination of disease is based on the comparison with traits or characteristics considered as "normal," "ideal," or "healthy." In other words: contrary to its own professional self-understanding, medical genetics cannot avoid, through the labeling of genetic diseases, making fundamental value judgments in regard to health and illness. Molecular genetics can indeed

> describe the human genome in its individual components in detail. In this process one can undeniably discover at most variety, yet never disease. Likewise, the numerous attempts to search for an all-embracing criterion of disease, by means of models of damage, of de-railing, of norm deviation, of imbalance, of missing harmony, of species-typical function, or some similar basis, do not change anything in the fundamental difficulties. For in these cases a criterion has been long applied in the above-mentioned models of deviation which cannot be simply derived from a description of nature. (Wiesing 1998: 96; cf. Koch 1999: 191–192)[21]

In this respect, there is no scientifically neutral description; rather, the definition of disease itself is a value-laden process implying decisions regarding action. A concept of disease that fails to take into account this normative dimension runs the risk of masking fundamental and inevitable moral decisions.[22] Urban Wiesing correctly points out that a central criterion for the existence of a disease lies in a subjective state of being unwell and an individual feeling of helplessness. Even if this "self-assessment" can only be invoked with some qualification and limits, it remains indispensable: "Otherwise one would label a person as 'ill' and his or her condition as needing change, while the person does not see it this way and does not want to be changed." (Wiesing 1998: 97) The structural problem of genetic medicine probably consists of the fact that, contrary to the principle of patient autonomy, it does not refer at all to the self-assessment of those affected but instead is oriented toward a seemingly objective conception of disease founded solely on molecular knowledge: "The critical factor for medical action is not the self-interpretation of a person, but a certain constellation of genes. And since self-interpretation and gene constellation by no means always correlate with one another, a certain risk exists here." (Ibid.: 97)

The philosopher of science Philip Kitcher has also drawn attention to the "hidden" normative foundations underlying genetic models of explanation. Kitcher shows that the assumption that genes exist "for" human diseases, behavioral characteristics, and traits always presupposes "some idea of *standard* genetic backgrounds and *standard* environments" (1996: 246; emphasis TL). These standard environments, which can be of a social or biological type, are responsible for whether and in what form a certain "genetic deviation" can actually lead to a change in the corresponding trait (on this see also Conrad 1999: 234). Kitcher not only emphasizes the need to integrate normative considerations into genetic explanations, but also indicates a further problem within the "molecular medicine" approach: the tendency to disregard or even systematically exclude the contribution of environmental factors to the formation of diseases.

Research findings in genomics and molecular biology indicate the limits of simple accounts linking phenotype and genotype in a unidirectional and linear way. Metaphors associated with the idea of genes as a blueprint for life—the Book of Life, the code of codes, the Holy Grail, and so on—are now rarely used to describe biological development. At the same time, there is more and more evidence that the "function" of a gene is not an inherent, isolated, and stable property but depends to a great extent on cellular and extra-cellular contexts and constellations. After the draft mapping of the human genome was published in 2001 and only around 25,000 genes were found (instead of the 150,000 earlier estimates had predicted), the idea of a more or less straightforward relation between individual genes and behavioral traits became less and less convincing. Furthermore, genome mapping brought to light that human DNA is much closer to that of other living organisms that expected (e.g. human beings share 98% of their genes with chimpanzees). Finally, findings in genomic research indicated that the non-coding parts of the DNA (once called "junk DNA") are in fact implicated in gene expression and regulation (Lock 2005: S47, S49). As a result, the reductionist concept of genetic regulation that had reigned in the past finally seemed to have been overcome by a more complex understanding of biological development. In this perspective, genes are more accurately conceived of as actors in an extraordinary complex network and not as determining agents (see Strohman 1997; Keller 2000; Lewontin 2001; Moss 2003; Lock 2005; The-Encode-Project-Consortium 2007).

It is tempting to interpret this transformation in genetic discourse as the well-known story of scientific progress that brings about a more accurate, complex biological understanding after necessarily simplistic scientific beginnings. Following Brian Wynne (2005), I argue that this familiar narrative is misleading and misplaced. In fact, it presents a rather reductionist account of how complexity is defined and targeted in contemporary genomics. Wynne analyzes different approaches to account for complexity in recent genomic research, and examines the different research agendas and methodologies that follow from these conceptions. He shows that the two dominant accounts simultaneously embrace and reduce complexity. Although they explicitly oppose genetic determinism, they nevertheless endorse a form of "neoreductionism" (Lock 2005: S48) that presents a very selective and restricted account of gene–environment interactions.

The first approach acknowledges the limits of the older paradigm of genetic causation and proposes a focus on whole genomes instead of individual genes to explain the variety of observable traits. This perspective stresses the importance of specific combinations and sequences of genes to account for multiplicity of phenotypes. However, this strategy relies on the very same linear and mono-causal mode of thinking that characterized molecular biology in the past. Shifting from individual genes to the whole genome as the determining singular agent does not overcome but rather restores genetic determinism. Alongside this genomic reductionism, more sophisticated understandings have emerged in rapidly developing fields like systems biology and epigenetics. This second approach to complexity focuses on interactions between whole-cell and whole-organism processes, on the one hand, and protein production and gene function, on the other. Here molecular processes are understood to be an integral part of a more comprehensive biological network characterized by mutual interactions and influences, and not simply an independent determinative agent. Within this complexity paradigm, it is neither individual genes nor the genome as a whole that account for biological processes; rather, genetic processes are analytically combined with and dependent on other levels of biological organization (Wynne 2005: 72).

However, even in this second approach there are important limitations to be noted that seriously restrict the perspective on complexity. It does not break with the techno-instrumental commitments of previous genetic explanations but continues them. Wynne correctly observes that biological complexity is framed and tamed by some important pre-analytical choices and guiding principles: "[B]iological complexity is defined within the parameters of predictability of selected phenotype properties, for the assumed purposes of drug therapy. This epistemic culture does not encompass the questions about what is not deterministically predictable, or is uncontrolled; it excludes them." (Ibid.: 77) Thus, ultimately, unpredictable effects or uncontrollable events may be unaccountably externalized onto the environment. As a result, virtually everything external to the material body remains black-boxed. Complexity is conceived as something that must be mastered. Everything that does not correspond to this instrumentalist vision is negated or ignored. As Wynne points out, reductionism is not a scientific problem, a fault to be rectified, but rather a legitimizing motor, something that allows for scientific authority and its predictive power.[23] It corresponds to a scientific culture that is more and more shaped by commercial imperatives and political interests. Reductionism is "actually systematic negation of any epistemic indications of non-control and non-knowledge beyond residual imprecision, which, it is assumed, can be domesticated" in the future (ibid.: 84). It follows that a different scientific culture rooted in complexity as recognition of limits to deterministic predictive control is effectively pre-empted (see also Huang 2000; Hedgecoe 2001; Lock 2005).

We are thus confronted with a contradictory picture. Although genomic research embraces complexity, opting for new interactionist models and integrative approaches that increasingly pay attention to feedback loops, multi-causality and non-linear trajectories, it is simultaneously unable to address complexity, because to do so would threaten the epistemic culture that demands control and predictability. Furthermore, allowing for complexity would also undermine the public authority and legitimacy of scientific expertise and the well-established view of the citizens as incapable of dealing with complexity. To allow for ignorance and lack of control weakens the role of science in translating complex issues to the larger public, and casts a shadow on the promises of future drugs and therapeutic options generated from a solid knowledge of the interactions between genetic processes and phenotypic "outcomes" (ibid.: 84–5).[24]

Indeed, we may note that a reductionist account of complexity lies at the heart of the contemporary vision of genomic medicine which systematically negates the insight that molecular processes are shaped by biological and non-biological environments. The appeal to complexity here reconnects with the promise of individualized medicine: a medicine tailored to one's genetic profile, which is conceived of as a more or less stable and independent structural entity. The rise of the concept of "genetic susceptibility" is quite telling in this respect. It replaces the older idea that there is a "gene for" a particular disorder by a search for genomic variations that might increase the probability of disease development:

> [W]hat susceptibility promises, in the age of genomics, is more than risk assessment and risk management—more . . . than intervention based on a correlation between factors such as age, weight, or diet whose link to the disease process may be unknown or distant. Susceptibility, it is claimed, is something that can be defined at the level of the individual body itself—a variation within the sequence of DNA bases

> in an individual's genomes that predispose that person to the development of a particular disease or disorder. (Rose 2007: 87; 84–95)

But this hope might be misguided, because it probably relies on an erroneous concept of biological regulation and genetic causation. It has to be replaced by a more integral approach that takes into account the fact that genes are not autonomous determinative agents but dependent on a complex regulatory network. If we concede that environmental, social, psychological, and biological factors may markedly alter the ways in which genes function, we end up with a cautious account that contrasts with the optimistic scenario of genomic medicine. A given genetic composition could give rise to a variety of biological outcomes, contingent on a range of further biological and non-biological factors. In this light, in the light of a more complex account of complexity, the promise of predicting disease-risk from genetic testing may be a false dream.[25]

Finally, we also have to note a fundamental paradox that characterizes the idea of genetic medicine. By promising the extensive diagnosis, prevention, and therapy of diseases that are genetically (co-)determined, it simultaneously expands the definition of disease to conditions and norm variations that up until now have not been considered as illnesses, and thus regards healthy persons as "genetically ill." The repeatedly proclaimed "victory" over disease actually renews and multiplies what it attempts to eradicate.[26]

2 Genetic Discrimination

Empirical Evidence and Regulatory Responses

The shifting understanding of disease and the increasing importance of genetic information for health care and clinical research have repercussions that far exceed the domain of medicine. The promising vision of "molecular medicine" casts the sinister shadow of new types of stigmatization, exclusion, and disadvantaging directed at individuals and families regarded and treated as "genetically ill." Some of these adverse effects have been debated for about twenty years under the heading of "genetic discrimination." The academic literature, the media, and the relevant legislation use this term to refer to the unjustified differential treatment of human beings with a (presumed) increased risk of being affected by a genetic disease. Generally, a strict distinction is made between genetic discrimination and discrimination based on disability and (chronic) illness. Whereas the latter relates to phenotypic factors, that is to say, to observable characteristics, the former is based on the genotype, in other words, the genetic make-up of the individual. Therefore, genetic discrimination occurs when an individual or that individual's relatives are unfairly treated differently because of actual or perceived deviances from the "normal" genome. This definition of the concept figured in the first empirical study pertaining to this subject in the early 1990s:

> [G]enetic discrimination is defined as discrimination against an individual or against members of that individual's family solely because of real or perceived differences from the 'normal' genome of that individual. Genetic discrimination is distinguished from discrimination based on disabilities caused by altered genes by excluding, from the former category, those instances of discrimination against an individual who at the time of the discriminatory act was affected by the genetic disease. (Billings et al. 1992: 477; see also the almost identical definition in Natowicz, Alper, and Alper 1992: 466)

Although the term genetic discrimination may be fairly new, the practices it describes are certainly not (Holtzman and Rothstein 1992). The history of genetics is closely linked with eugenic practices directed at individuals deemed "genetically inferior" because of their family background. The spectrum of these interventions ranged from forms of health education to encourage individual responsibility for reproductive conduct to segregation and compulsory sterilization of those considered "feeble-minded" or "genetically defective" In many European countries, as well as in the United States. The most radical manifestation of eugenics was the active killing of disabled individuals to ensure that they did not pass on their defects to the next generation in the era of National Socialism (Adams 1990; Weingart, Kroll, and Bayertz 1992; Kevles 1995; Dowbiggin 2008; Bashford and Levine 2010). The negative classification of individuals as "genetically disabled" by no means ended after World War II. In some countries, sterilization programs continued after the war, and eugenic arguments still have currency in both scholarly and public discourse even today (Duster 1990; Kühl 1994; Roll-Hansen and Broberg 1996; Paul 1998; Koch 2000; Rembis 2009).[1]

The history of eugenics shows that (predictive) genetic testing is not necessarily required for discriminatory practices to take place. Apart from social prejudices and stereotypes, adverse treatment can also be based on an individual's medical reports or family history. What is new, however, and this is what sets genetic discrimination apart from its precursor phenomena, is that the development of predictive testing facilitates the identification of individuals at risk of developing a genetic disease, without any need to refer to family history or wait for an actual manifestation of the illness. This is supplemented by a further systematic difference. Whereas the eugenic practices of the past focused on distinctive subjects and population groups marked by more or less observable characteristics, genetic discrimination is grounded in a non-manifest genotype that is conceived as "different" or "deficient." Those affected by genetic discrimination are not (yet) ill; rather, they are the "asymptomatic ill" (Billings et al. 1992: 479) who may in the future, or possibly never, suffer from a disease for which a genetic component is made causally responsible.[2]

In principle, five categories of the "asymptomatic ill" can be distinguished that might be affected by practices of genetic discrimination (cf. Taylor et al. 2004: 228; Barash 2000: 217):

1. Pre-symptomatic individuals who have tested positive for a highly penetrant condition—such as Huntington's disease—which in all likelihood they will develop later in life
2. Predisposed individuals who have been diagnosed with a genetic disposition and who have a significantly higher probability of developing a multi-factorial disease, although without certainty (e.g. hereditary breast cancer)
3. Individuals with a fully treatable or preventable genetic condition (e.g. hereditary hemochromatosis)

4. Individuals who are "carriers" of mutations in recessive genes which may be passed on to their children, but who will themselves never develop the disorder (e.g. carriers of the CFTR gene for cystic fibrosis)
5. Individuals who have been diagnosed with a genetic illness, but who fail to show any physical symptoms (e.g. some of those diagnosed with Gaucher's disease)

The idea of an independent form of discrimination which is based on genetic factors, and which is to be categorically distinguished from discriminatory practices based on phenotypic characteristics, has dominated both theoretical debates and empirical research on the (negative) social implications of genetic knowledge. This chapter will summarize research findings from empirical studies conducted on experiences of genetic discrimination so far and outline the current status of legal regulations pertaining to genetic discrimination, focusing on the regulatory frameworks in the United States and Germany.

REVIEW OF EMPIRICAL RESEARCH

Since the early 1990s, several empirical studies have investigated the phenomenon of genetic discrimination.[3] The crucial impetus was provided by a group of U.S. intellectuals who have been researching the social and legal implications of scientific progress in genetics for over twenty years. Initially located in and around the Boston area, the members of the Genetic Screening Study Group now live and work all across the United States. The group contains both natural and social scientists, but also philosophers, physicians, lawyers, and representatives of genetic support groups. The founding members include the chemist Joseph S. Alper, the molecular biologist Jon Beckwith, and the jurist Lisa N. Geller; the physician Paul Billings, the sociologist Peter Conrad, and many others later joined them. The first two studies to attempt to determine the extent and forms of genetic discrimination in the United States were planned and conducted by the Genetic Screening Study Group.[4]

The first was an explorative investigation carried out in the early 1990s under the direction of Paul Billings (Billings et al. 1992).[5] Billings and his team published an appeal in the *American Journal of Human Genetics* asking physicians and genetic counselors to inform them about cases in which individuals had been discriminated against on account of their genetic make-up. A similar appeal was sent to physicians active in clinical genetics and to self-help organizations for those affected by genetic diseases. Billings and his co-authors received forty-two responses from across the United States and Canada, although thirteen (31 percent) of these could not be considered because it was unclear in these cases whether the adverse treatment was in fact caused by the suspected or confirmed genetic characteristic. A total of twenty-nine responses remained after this critical review, supplying evidence of forty-one cases of genetic discrimination. With only two exceptions, all of these concerned the insurance sector (health, life, and car insurance) or employment (hiring, dismissal, training, and promotion within companies). The investigators also established that the reported experiences showed certain similarities and could be divided into three thematic blocks or groups of affected individuals.

The first group of cases comprised individuals without clinical symptoms, who were perceived as different only due to their possession of a genotype deviating from a "normal" genetic make-up. In two cases, one concerning an individual with hereditary hemochromatosis[6] and one related to phenylketonuria (PKU), the manifestation of the disease was hindered by early treatment or a special diet. Another case concerned a person carrying a mutation for Gaucher's disease[7] who would never fall ill (because it is a recessive disorder, both parents must pass on the gene for the child to be affected). Despite the fact that all these individuals were healthy, the former were denied health insurance and the latter was refused employment with a government authority on grounds of his genetic status. Based on these case examples, Billings and his co-authors concluded that the dissemination of genetic tests leads to the emergence of a new category of individuals:

> As large numbers of individuals submit to or are coerced into testing in order to obtain employment or insurance coverage, a new social class and category—the "asymptomatic ill"—may be constructed. Although they are healthy, persons in this new group may find that they are treated as if they were disabled or chronically ill by various institutions of our society. (1992: 479)

The second thematic block addresses the problem of genetic variety. Even in so-called hereditary diseases there is rarely a simple and straightforward causal connection between genetic traits and disease. Thus, some individuals diagnosed with the inherited metabolic disorder cystic fibrosis exhibit severe symptoms, whereas others show no or only minimal signs of health impairment. One and the same genetic mutation can have very different clinical effects and, vice versa, a single disease may be caused by a variety of genetic mutations. Billings and his colleagues describe two cases involving individuals with mild symptoms of Charcot-Marie-Tooth disease, a muscular disorder which can lead to severe movement impairment but which is never fatal. Nevertheless, the affected individuals were denied both life and car insurance coverage in one case and employment in the other. In contrast to the first category, these persons were actually affected by a disease, although one whose clinical symptoms were not particularly manifest and hardly impaired them in their everyday lives. The insurance companies and the employers based their decisions solely on genetic knowledge and disregarded the actual repercussions the disease had on each individual. To them a particular genotype was tantamount to the presence of a severe disease for which no effective therapeutic options exist.

The third group of cases involved individuals who at that moment were healthy, but who were at risk of being affected by a genetic disease in the future. Adoption agencies, for example, refused to assist individuals who were at risk of Huntington's disease in adopting a child. One of them vividly described how he felt stigmatized by this exclusion criterion:

> We understand the right to choose the *best* 50 couples out of some 500 applicants per year for placement. Availability of children is certainly limited. And yet, should I be judged by a disease that I am only at risk for and that may not develop for some years to come? Does this make me different from anyone with diabetes or cancer, for example, in their ancestry? (quoted in Billings et al. 1992: 480; emphasis in original)

The stance taken by the adoption agency shows how, in social and institutional practice, genetic risks are treated differently from other health risks.[8] Another case concerned a family with a child living with cystic fibrosis who received care benefits from their health insurance. When the mother became pregnant a second time, prenatal tests were conducted which verified that the fetus carried two copies of the mutated gene, which would most certainly cause the disease. When the family decided not to interrupt the pregnancy, the insurance company resolved to reduce or discontinue payments, both for the unborn child and for the child who was already ill. Only under threat of legal action was this decision rescinded. In the authors' view, these incidences reveal the hegemony of the "myth of genetic perfection" which ultimately points to a eugenic prejudice: "The agencies assume that the best possible family is the one least likely to face medical adversity—the 'perfect' family with a disease-free genome." (Billings et al. 1992: 480)

The study conducted by Billings and his colleagues met with a strong reaction in the United States, and stimulated a lively debate on the adverse social implications of increasing genetic knowledge and to what extent third parties might use genetic data. Based on specific case examples, it confirmed the existence of a particular form of discrimination that individuals are subjected to when they are treated as "asymptomatically ill" based on genetic characteristics. Significant and groundbreaking though this pilot study was, its most obvious shortcoming was its limited database. This made it unsuited to the drawing of conclusions about the quantitative dimension of genetic discrimination.[9]

This problem was corrected by the first comprehensive study of the forms of genetic discrimination, likewise instigated by the Genetic Screening Study Group and published in 1996 under the direction of Lisa N. Geller (Geller et al. 1996). The approach was similar to the earlier study by Billings and his colleagues. Again, the focus was on collecting case reports by surveying those affected. This was carried out in the United States between 1992 and 1993, with a considerable increase in the number of persons surveyed compared to the precursor study. In total, the authors sent out 27,790 questionnaires to individuals in whose families the following diseases had occurred: hemochromatosis, Huntington's disease, PKU, and mucopolysaccharidosis.[10] Contact to individuals possibly affected was made via their support groups; where questionnaire responses indicated cases of genetic discrimination, telephone interviews were arranged to substantiate the information. Of the questionnaires sent out, 917 were returned. Out of these, 455 respondents reported that they had had experiences involving genetic discrimination, whereas 437 said they had not experienced this (the remaining answers were inconclusive). After a critical examination of all responses, the authors selected 206 cases which corresponded to their definition of genetic discrimination and conducted extensive interviews with these respondents over the telephone. The results showed that discriminating practices occurred in a variety of different institutions. However—as before in the study of Billings et al.—there was one outstanding area of discrimination: the majority of cases, again, related to health and life insurance. The rest concerned the medical sector, adoption agencies, the military, educational institutions, blood banks, and employers.

As in the previous study, Geller et al. compiled cases that represented different forms of genetic discrimination, ranging from denial of insurance coverage and exclusion from blood donations to teachers' refusal to promote students to the next class because they interpreted insufficient school performance as the first signs of a genetic disease (ibid.: 75–78). In addition to recording practices of (alleged) genetic discrimination, the study also aimed to identify the strategies with which those individuals and families meet the threat of stigmatization, disadvantaging, and exclusion. These encompass, among others, providing false information when asked about their family history, taking genetic tests anonymously or paying for them in cash so that insurance companies would not find out about them, or purchasing insurance policies *prior* to a genetic test. The study also determined, however, that there is still a lack of knowledge on the part of those affected by genetic discrimination on how to legally challenge these practices. The authors concluded that genetic discrimination must be regarded as a social fact already effective in a variety of social fields, with significant repercussions on those individuals and families affected by it (ibid.: 78–83; cf. also Geller et al. 2002).[11]

The major achievement of these two pioneer works from the Genetic Screening Study Group was to sensitize the scientific community to the adverse side effects of growing genetic knowledge. They were followed by further empirical studies on genetic discrimination. The same year in which Geller and her colleagues brought out their work also saw the publication of a study by E. Virginia Lapham, Chahira Kozma, and Joan O. Weiss (1996). The authors conducted a survey among 332 members of families in which one or more genetic diseases had been diagnosed about their experiences concerning genetic discrimination. These individuals were contacted via self-help initiatives associated in the Alliance of Genetic Support Groups. Lapham, Kozma and Weiss conducted in-depth telephone interviews with the majority of participants, and others filled in questionnaires. Respondents came from all across the United States and represented a wide spectrum of genetic diseases. Twenty-five percent reported having been denied life insurance coverage, either for themselves or a family member, because of their genetic characteristics; 22 percent were turned down by health insurance companies on the same grounds; and 13 percent reported workplace discrimination (either with regard to hiring or dismissal) (Lapham, Kozma, and Weiss 1996).[12]

Mark A. Hall and Stephen S. Rich directed a comparative case study focused on seven U.S. states with and without legislation against genetic discrimination, and published the results in 2000. These authors addressed the question whether, and in what way, health insurance companies discriminate against applicants and clients based on their genetic properties. For this purpose, the authors conducted interviews with representatives from the political and administrative sectors as well as with insurance companies, genetic counselors, patient advocates, and the medical director of a genetic diagnostics firm. In contrast to the aforementioned studies, they arrived at the conclusion that neither at the time of investigation nor in the past did any noteworthy forms of genetic discrimination occur. Hall and Rich reported that they were able to document only two cases of genetic discrimination. In both of them, the insurance company initially refused to cover the costs of a genetic consultation (one case concerned the genetic test for Huntington's disease, the other a test for a hereditary form of cancer) and after that raised the premium, suspecting a genetic disorder without waiting for the actual test results. All in all the authors claimed that the problem of genetic discrimination is frequently exaggerated, and that legal measures are effective primarily on a symbolic level: "Multiple, independent sources refuted, or were unable to document, any substantial level of genetic discrimination by health insurers, either before or after the legal prohibitions were enacted, or in states with or without these laws." (Hall and Rich 2000: 302)[13]

There are essentially three reasons that explain why the academic discourse surrounding genetic discrimination started comparatively early and had more momentum in the United States. First, the technological development is most advanced, and the use of genetic tests quite widespread in the United States. According to a survey by the *American Management Association* in 1999, 30 percent of medium-sized and large enterprises already expressed an interest in their employees' genetic data at that time; 7 percent drew on them when making decisions concerning hiring and promotion (Martindale 2001: 14; cf. also Pagnatarro 2001; Deering 2002). Second, the political and institutional context plays an important role. Unlike many other Western countries, the United States does not yet have a public health insurance system, which makes risk differentiation based on genetic anomalies an existential matter. In extreme cases, applicants are rejected and thus left without any health insurance at all. Third, there was a far more intense public debate on genetic discrimination in the United States as compared to most European states due to stronger patient advocacy groups that addressed the (adverse) social implications of genetic knowledge in their work (Heath, Rapp, and Taussig 2004; Rose and Novas 2005). In contrast, in Europe the public debate on genetic discrimination was mostly confined to the discussion of regulatory proposals put forward by policy actors and state administrations.[14]

However, practices of genetic discrimination are not limited to the United States. Based upon a survey among individuals diagnosed with a variety of genetic diseases, Lawrence Low, Suzanne King, and Tom Wilkie (1998) reported discriminatory practices among British insurers. The authors sent out 7,000 questionnaires to members of self help initiatives for individuals living with genetic diseases (just over 1,000 forms were sent to a control group made up of a representative sample of the population). Altogether, 33.4 percent of respondents reported problems applying for life insurance; in contrast, only 5 percent of applicants in the control group mentioned difficulties in acquiring life insurance. The survey also revealed that many cases of genetic discrimination are a result of insurers' or their representatives' lack of expert knowledge regarding causes and syndromes of diseases.[15] The watchdog organization GeneWatch UK recorded cases of genetic discrimination in the workplace (Mayor 2003). The United Kingdom permits the use of genetic tests in pre-employment medical examinations. As a survey among employers revealed, 50 percent deem it appropriate to use genetic tests to identify individuals with an increased risk of disease (GeneWatch UK 2001, 2003).

At least two examples emerged in France where insurers attempted to discontinue policies on grounds of positive test results for Huntington's disease (Browaeys and Kaplan 2000: 2; cf. also Thébaud Mondy 1999). Several cases also occurred in Canada, where insurance companies rejected applicants because of their genetic characteristics or where insurance claims were denied because the applicant did not inform the insurer he/she carried a mutation for a genetic disease (Lemmens 2000: 352–354).[16] In Hong Kong, three young men who applied for posts with the fire brigade and the financial authority were turned down because of their family histories. In each case, one parent suffered from schizophrenia. The responsible public authorities refused to hire the applicants and based this decision on the (presumed) genetic component of the illness, claiming that the increased risk of disease made the applicants a serious danger to public safety. All three applicants appealed successfully against this decision (Wong and Lieh-Mak 2001). In Brazil, a volleyball player was excluded from the official team after a routine examination identified the sickle cell trait in her blood. As she was only a bearer of the trait but not affected by the disease, the exclusion was not supported by medical evidence (Guedes and Diniz 2007).

In Australia, initial case studies recorded a total of forty-eight cases of genetic discrimination, the majority of which related to the insurance industry and employment relations (Taylor 1998; Keays 2000; Barlow-Stewart and Keays 2001). The Genetic Discrimination Project, a comprehensive study on genetic discrimination in Australia, was undertaken from 2002 to 2005. Under the direction of Margaret F. Otlowski, the study contained various analytical dimensions and combined a survey of "consumers" (individuals who seek genetic counseling and members of self-help initiatives for genetic diseases) with a number of cases involving genetic discrimination in relation to employment or the insurance sector. In addition, it analyzed legal cases in which allegations focused on genetic discrimination. The last analytical step ("Verification Project") examined cases of alleged genetic discrimination (with the explicit consent of "consumers" and relevant third parties). Here, the aim was to reach a better understanding of individual experiences of genetic discrimination, the social and organizational contexts in which it takes place, and the patterns it follows (Otlowski, Taylor, and Barlow-Stewart 2002; Taylor et al. 2004).[17]

The individual projects are now complete, and publication of the research results is currently in progress (Otlowski et al. 2007; Barlow-Stewart et al. 2009). One of the studies was a large-scale survey of clinical genetics clients. The researchers conducted a survey by mailed questionnaire of past clients of clinical genetics and familial cancer services in Australia. They contacted 2,667 adults who had got in touch with clinical genetic services from 1998 to 2003 regarding predictive or pre-symptomatic testing for designated mature-onset conditions; 951 respondents (in total: 1,185) met the inclusion criteria for current asymptomatic status. Specific incidents of alleged negative treatment were reported by 10 percent (n=93) of respondents (most of them relating to neurological conditions and familial cancers). The incidents covered five domains: life insurance (42 percent), employment (5 percent), family (22 percent), social (11 percent), and health (20 percent). Incidents related to the family domain included

> negative treatment associated with a decision to undertake testing or a genetic test result (18 times), negative reactions/rejection from prospective family members regarding respondents' suitability for marriage or parenthood (3 times), advice/instruction not to reproduce (4 times) and disclosure of genetic information to a sibling without consent (n = 1). (Taylor et al. 2008: 26)

The study also provided evidence that a minority of respondents knew about the possibility of a complaint process. Only 15 percent of them knew where to report and challenge negative treatment because of genetic issues (ibid.: 27; see also Taylor 2011).[18]

The empirical studies on genetic discrimination conducted since the beginning of the 1990s have highlighted the complexities and the breadth of discriminatory experiences that relate to a perceived or factual knowledge of an individual genotype. Beyond the insurance and employment sector, the studies identified many other social settings and institutional domains where individuals at risk from a genetic disease and their families have been exposed to forms of negative treatment they consider discriminating on the basis of their genetic characteristics.

LEGAL AND POLICY RESPONSES TO GENETIC DISCRIMINATION: THE DIFFICULTIES OF REGULATION

The academic discourse and empirical studies dedicated to the issue of genetic discrimination have not gone unheeded. Since the early 1990s, a number of legislative initiatives and statements have been launched and issued by inter- and supranational organizations and commissions with the aim of protecting individuals against genetic discrimination. Thus, article 6 of the UNESCO *Declaration on the Human Genome and Human Rights* reads: "No one shall be subjected to discrimination based on genetic characteristics that is intended to infringe or has the effect of infringing human rights, fundamental freedoms and human dignity."[19] At the European level, explicit bans on discrimination based on genetic characteristics are contained within the *European Convention on Human Rights and Biomedicine* of the European Council (1997) (article 11) as well as in the *Charter of Fundamental Rights* of the EU (2000) (article 21).

Many countries around the world have enacted genetic non-discrimination regulations to restrict or ban the use of genetic information in employment and insurance (Joly, Braker, and Le Huynh 2010; see also Rothstein and Joly 2009; Knoppers, Godard, and Joly 2004). The laws are intended to protect individuals against unfair differential treatment based on genetic knowledge in the fields of employment and insurance, and to respond to the fears of individuals currently dissuaded from undergoing genetic testing in both clinical and research settings. Today, most European states have introduced regulations to ensure that no individual is discriminated against because of his or her genetic constitution. In contrast to the United States, European countries often provide some form of universal access to health care and accord a much more important role to the social dimension of insurance. Some countries have adopted a prohibitive approach banning the use of genetic test results by the insurance industry (e.g. Austria, Belgium, France), whereas others have opted for moratoria requiring the insurance industry neither to request genetic testing nor to use test results for insurance contracts for a certain period of time (United Kingdom, Finland, the Netherlands) (Van Hoyweghen and Horstman 2008; Joly, Braker, and Le Huynh 2010: 361–363; see also McGleenan and Wiesing 2000).[20]

Germany could be described as a "late developer" concerning genetic information regulation (Ireni-Saban 2010: 366). A law aiming to protect individuals against genetic discrimination was initially announced in 2002. The draft published in 2004 sought to establish a framework to regulate genetic testing for medical purposes, paternity testing and the use of genetic information by insurance companies and in employment. After several years of debate the Genetic Diagnostics Act (Gesetz über genetische Untersuchungen bei Menschen, GenDG) was finally passed by the German parliament and came into effect on February 1, 2010.[21] The focus of the Act is on the right to information and self-determination, with the aim of protecting individuals against abuse of their genetic data. Human dignity is emphasized as the normative foundation of the act (Backhaus 2012; Lander 2012).

The Act explicitly states (§4) that no person may be discriminated against or disadvantaged because of genetic characteristics. It generally prohibits insurers (§18) and employers (§19) from demanding a genetic examination or the results of any previously conducted genetic examination. The Act also underlines the obligation of medical professionals to inform individuals thoroughly about, for example, the risks of the examination and the patient's rights (§9). Under §10, genetic counseling is obligatory for the individual seeking genetic testing.

The Genetic Diagnostics Act focuses on the use of genetic information by employers and insurers and takes up some of the recommendations of the commission of enquiry of the German Bundestag (Deutscher Bundestag 2002). Insurance companies are prohibited from requesting clients to undergo genetic tests prior to the conclusion of insurance contracts. Furthermore, they must not even receive or use genetic information given to them voluntarily by the individual. However, the prohibition of use does not apply if the agreed benefits in life, disability, or pension insurance policies exceed 250,000 euros or 30,000 euros in annuities. This exception is justified with a declared necessity of protecting the insurer against adverse selection. Similarly, the intention with regard to employment relations is to prevent employers consulting genetic screening results when making their hiring decisions. But again, exceptional cases are acknowledged: genetic tests are to be permitted in the context of occupational medicine examinations to identify employees who are particularly at risk in "professions endangering their health" (Eberbach 2010; Kröger 2010).[22]

Although the aim of the Act is to protect against practices of genetic discrimination, critics have stressed regulatory deficits and shortcomings as well as normative contradictions which might be addressed in the course of further parliamentary consultations and in future clarifications by the Genetic Diagnostic Commission (Gendiagnostik-Kommission, GEKO).[23] For example, protection against discrimination in the workplace does not apply to civil servants of a German state. This question is of particular relevance, because applicants with genetic anomalies have had difficulties obtaining posts in public service in the past (see Chapter 4). Also, the Act does not regulate genetic research. Hence, it does not specify the purposes and conditions under which third parties should be allowed to access genetic data stored in "biobanks" for scientific or medical research (Eberbach 2010: 155). A further concern is that the GenDG provides no complete and coherent definition of a genetic test. In practice, it is often difficult to identify the borderline between predictive and diagnostic testing or between protein-based tests and genetic tests. Therefore, insurers are often unable to tell precisely which uses of predictive information the law covers (Backhaus 2012). And finally, in the context of immigration and family reunification important legal guarantees are inoperative. For example, immigrants cannot demand that their genetic data be destroyed, and their data might be used for criminal prosecution purposes (cf. Genetic Diagnostics Act §17, 8). Critical attention has been drawn to the fact that the Genetic Diagnostics Act implements a double standard, according a special status to immigrants by refusing them the right to informational self-determination while at the same time extending the legal guarantees for native citizens. As a result, DNA testing in the context of immigration has become a hot topic in public debates in Germany with a lot of media coverage (e.g. Gaserow 2007; Schmidt 2007; Funk 2009; Heinemann and Lemke 2013).[24]

In the United States, nearly every state has passed legislation to protect individuals against genetic discrimination in the fields of employment and insurance, but until recently there was no federal regulation.[25] Although genetic non-discrimination legislation was introduced in 1995 during the 104th Congress, it took thirteen years of intense debate before the Genetic Information Non-Discrimination Act (GINA) passed the House of Representatives and the Senate. On May 21, 2008, President George W. Bush signed GINA into law.

Like its German counterpart, GINA protects individuals from being discriminated against in health insurance and in employment via the use of genetic information. The bill applies a broad definition of genetic information that not only includes results of genetic testing of an individual and his or her family members, but also information about family medical history. The latter was included in the definition because practices of genetic discrimination have often referred to family history to determine an increased risk of developing a disease or disorder in the future. The law took full effect in November 2009. It was supported by many medical and scientific organizations (e.g. the American Society of Human Genetics and the American Medical Association) and support groups like the Coalition for Genetic Fairness, and applauded by Senator Ted Kennedy as "the first civil rights bill of the new century of the life sciences" (quoted in Coalition for Genetic Fairness 2008: 3; Slaughter 2008; Roberts 2010).[26]

In Title I of the bill, GINA defines unlawful practices for health insurers in the use of genetic information. In principle, it prohibits health insurers from requiring individuals to provide their genetic information or the genetic information of a family member to the insurer for eligibility, coverage, underwriting, or premium-setting decisions. Also, health insurers may not request or require that an individual or an individual's family member undergo a genetic test. The protections in employment settings are specified in Title II of the bill. GINA prohibits employers from using genetic information in making decisions regarding hiring, promotion, terms or conditions, privileges of employment, compensation, or termination. An employer, employment agency, labor organization, or training program may not limit, segregate, or classify an employee or member on the basis of genetic information. Only in some very narrow and well-defined circumstances may these actors and agencies have access to genetic information on their employees or members (e.g. when the information is publicly available or has been inadvertently provided). Also, they are requested to treat the genetic information they possess about an individual as a confidential medical record and maintain it in separate forms and in separate files (Slaughter 2008; Tan 2009; Roberts 2010).[27]

GINA is the first federal law in the United States to address the problem of genetic discrimination in employment and health insurance. However, commentators have pointed to some serious shortcomings and loopholes. Although the regulations in health insurance are rather strict, allowing only for rare exceptions, the protections provided by the law do not include life, disability, or long-term-care insurance and other possible uses of genetic information in the insurance industry (Rothstein 2008a, 2008b). Also, GINA does not interfere with the employer's obligation to provide for a safe and healthy work environment—which might require the collection of genetic data for genetic monitoring programs (e.g. of the biological effects on employees of toxic substances in the workplace) (Hudson 2007; Coalition for Genetic Fairness 2008: 10). It also has to be noted that the health insurance provisions of GINA do not cover some groups. For example, the bill does not apply to members of the U.S. military, federal employees obtaining healthcare through the Federal Employees Health Benefits Plans, or employers with fewer than fifteen employees. There is another group that remains outside of the legal protection of the law. Whereas those individuals conceived of as asymptomatically ill based on genetic characteristics are protected by the GINA and individuals with manifested diseases are covered by the Americans with Disabilities Act (ADA), individuals displaying mild symptoms of a disease or temporary conditions are not protected either by the ADA or by GINA. Furthermore, the legal protection provided by the two laws does not cover individuals "with a genetically based, biologically determinable difference beyond genotypic variation but short of phenotypic variation" (Rothstein 2008a: 839; see also Kang 2011). Hence, Rothstein postulates the emergence of a new "uncertainty . . . in the middle" (ibid.) and calls for an amendment of the non-discrimination legislation that eliminates the current gaps in coverage. According to Rothstein, this problem is based on a "scientifically dubious dichotomy between genetic and non-genetic information, tests, or disorders"[28] (ibid.; see also Korobkin and Rajkumar 2008).

There are in principle two legal-political options currently applied to govern the use of genetic information, especially in the employment and insurance sector (see Gerards, Heringa, and Janssen 2005). The first places special emphasis on the protection of genetic data and focuses on the right to genetic privacy. This form of legislation proceeds from the principle that every citizen should be in control of his or her genetic data, which are not to be passed on to non-medical parties without the explicit and informed prior consent of the individual concerned.[29] The legal guarantee of data protection alone, however, is insufficient. If the refusal to reveal genetic information is expected to entail severe social disadvantages (if, for example, insurance policies are denied or career goals threatened), individuals could *de facto* feel forced to forego exercising their rights.

In an effort to avoid this, a second regulatory instrument centered on the principle of non-discrimination is often employed. Rather than prohibiting the collection and storage of genetic information by institutional actors, non-discrimination law aims at curbing and preventing the use of such information. Both these legal-political strategies have their advantages and disadvantages and are often combined in legal practice. While the strategy of data protection limits access to genetic information, the non-discrimination principle prohibits its inappropriate use for organizational decision-making processes:

> [T]he nondiscrimination principle requires that the individual must not only show that he has been disadvantaged because of his personal genetic data, but also that other individuals or groups (lacking the relevant genetic defect) have been treated more favorably. This comparative aspect of the nondiscrimination principle seems to make it somewhat less useful and valuable to the individual than the privacy approach, which does not require a comparison to other groups or individuals: to establish a violation of genetic privacy . . . On the other hand, an advantage of the discrimination approach may be that the individual does not have to show that the insurer or employer has actually invaded his personal privacy by using genetic information. (Gerards, Heringa, and Janssen 2005: 24)

In the context of insurance underwriting, it is possible to discern different political and legal approaches combining the two strategies to regulate the use of genetic information. Internationally, the main approaches favored by policymakers to solve the problem of how to use genetic knowledge in insurance are identified in the literature as the following: (1) a "'fair limits' approach" that permits "insurers to access and to use genetic information for insurance underwriting only for policies above a legislatively predetermined amount of money"; (2) a "human rights approach" "stemming from the field of international human rights law aiming to prevent the discrimination of individuals on the basis of genetic information"; (3) a "moratorium approach" that is defined as "a voluntary agreement by a group of insurers (often through an official representative organization), to neither request genetic testing of insurance applicants nor to use genetic test results for a certain period of time"; (4) a "prohibitive approach" "aimed at specifically prohibiting access to or use of genetic information by the insurance industry"; (5) a "'rational discrimination' approach" "permitting the use of genetic information for insurance underwriting only after it has been deemed scientifically valid and clinically significant by an independent expert scientific panel"; and (6) a "status quo approach" consisting of a "wait-and-see default approach to the use of genetic information by insurers. This approach allows insurers to develop their own rules pertaining to the use of genetic information in connection with insurance contracts without intervention from the government" (Rothstein and Joly 2009: 138–139; see also Knoppers, Godard, and Joly 2004).[30]

Whatever the regulatory framework may be, in legal practice, any approach addressing genetic discrimination faces two essential problems. One concerns the extent of the regulation, the other its epistemological basis. With regard to the first, it needs to be clarified whether every form of differential treatment is affected by the regulations against genetic discrimination or merely those considered "unfair" and unjustified. Is every form of differential treatment based on genetic information to be regarded as discriminatory and thus illegitimate, or only those for which there is no justification (whatever this might mean)? Second, there is a need to clarify what precisely is to be understood as "genetic information." Concerning this question, some legal texts operate with a narrow sense of the term, in which case the regulation applies solely to the use of data extracted from genetic tests; others, however, propose a wider definition, according to which even family history details are not to be drawn on for insurance contracts, employment relations, and so on.

A further controversy focuses on the question of whether legal protection against genetic discrimination requires specific legislation or whether it is already covered by existing regulations. For instance, the question needs to be addressed of whether discrimination against the "asymptomatic ill" based on genetic characteristics represents a particular form of discrimination against people with disabilities. From this perspective, no specific legislation would be necessary to protect individuals with genetic anomalies as long as regulations exist for protection against ableism.[31] Especially in the United States, scholars have frequently pointed out that the definition of disability contained in the ADA is so broad that it embraces individuals with genetic anomalies (cf. Gin 1997; Silvers and Stein 2002). With reference to the German context, legal scholars Rainer Paslack and Jürgen Simon have openly expressed fundamental doubts about the need for a genetic non-discrimination act. For them, special legal norms that focus on practices of genetic discrimination are simply redundant. One can only speak, they argue, of discrimination against a person in the sense of injury to their dignity due to their presumed genetic constitution in the strict sense as being equivalent to acts of racial or sexual discrimination. According to these authors, because German constitutional law provides sufficient safeguards for the protection of the individual, special regulations for genetic discrimination are simply superfluous (2005: 148–149).[32]

However, one can doubt whether constitutional law suffices to regulate such a complex legal subject thoroughly and permanently. It still has to be determined which test procedure applied to which individual, to what end, and using which information shall be allowed. This problem is strikingly illustrated by the discussion that arose in the wake of the genetic screening for hereditary haemochromatosis offered by the German public health insurer Kaufmännische Krankenkasse (KKH) to its policyholders in cooperation with scientists from the Medizinische Hochschule Hannover. From 2001 to 2004, nearly 6,000 policyholders inquired about the genetic test, and 3,961 ultimately provided blood samples. In November 2004, upon completion of this "model test," the health insurer summarized the results in a press release stressing that the first large-scale genetic screening by an insurance company in Germany brought to light the fact "that preventive genetic diagnostics facilitates an enhanced quality of life and the saving of millions of euros" (quoted in Steindor 2005: 12; Georgescu and Vollborn 2004).

However, this assessment did not go unchallenged. On the one hand, both the scientific validity and the medical significance of the screening in terms of its predictive capacity for hemochromatosis are fiercely debated. In 2010, the European Association for the Study of the Liver issued clinical practice guidelines to assist physicians, health care providers, and patients in clinical decision making for hereditary hemochromatosis. The guidelines explicitly refrained from recommending genetic screening of the general population because "disease penetrance is low and only in few C282Y homozygotes will iron overload progress" (European Association for the Study of the Liver 2010: 8; see also Fleming and Ponka 2012). On the other hand, the number of samples delivered and stored for this enterprise gave rise to concerns regarding data protection. After the pilot study the German Federal Commissioner for Data Protection at that time, Peter Schaar, called for clear legal regulation with regard to genetic screenings. He argued that there was a need to ensure that only the tested person him-/herself would receive the test results. The diagnosis should be disclosed neither to the public or private health insurer nor to the employer—not even with the prior consent of the tested person. Furthermore, it was necessary to specify that those who administered the genetic test could not assign the results to any specific individual. Finally, he considered it necessary to destroy the collected samples after completion of the genetic tests (Siems 2004).

This example shows the complexities of the scientific debates and regulatory issues at stake, stretching from constitutional law via non-discrimination regulations to data protection provisions. Concerning the assessment of genetic discrimination and its regulation, there are a number of open questions and controversial debates. In academic discourse, there is still no agreement about how widespread practices of genetic discrimination actually are or about whether genetics-specific regulation is necessary and helpful in this domain. Whereas some scientific observers maintain that genetic discrimination is a crucial social problem requiring legal safeguards (Geller et al. 2002), others conclude that the phenomenon is no more than a marginal one and legal regulation is superfluous or even damaging (Hall and Rich 2000).

The scientific discourse surrounding the necessity of a specific statutory protection against genetic discrimination may be controversial, but the fear of genetic discrimination is a fact. This is indicated by evidence provided by a series of empirical studies in different countries. In the largest survey to date, over 86,000 individuals from different states of the United States and Canada were questioned about their attitudes toward genetic discrimination. Each respondent participated in a screening scheme for hemochromatosis. The study provided clear evidence that a large number of respondents fear the negative effects of genetic discrimination: 40 percent of participants agreed with the statement, "Genetic testing is not a good idea because you might have trouble getting or keeping your insurance." Because the survey participants had already given their consent to take part in the screening scheme, and because it can therefore be assumed that these individuals are basically in favor of genetic tests, skepticism in the general population is likely to be much stronger (Hall et al. 2005; see also Matloff, Brierley, and Shappell 1999; Matloff et al. 2000).

Concerns relating to the uptake of genetic testing by those affected by particular diseases are also documented by an empirical study conducted in the Netherlands. Between 2004 and 2008, members of six Dutch extended families living with (the risk of) hypertrophic cardiomyopathy (HCM) were followed. In interviews on the decision to take up genetic testing for HCM, family members often articulated fears of possible discrimination in relation to insurance and employment which resulted in postponing genetic testing, feelings of regret after undergoing genetic testing, or the wish not to know the test results (Geelen et al. 2012; see also Imke et al. 2010). A survey conducted in the United Kingdom produced similar findings. It revealed that every British breast cancer center reported women who, fearing problems with their insurer, turned down the offer of a genetic screening test. In 40 percent of centers, this refusal was induced by a fear of workplace discrimination (Morrison 2005: 878). Hence, an important consequence of these fears is either people's refusal to avail themselves of genetic tests which might potentially be medically beneficial or, in an effort to remain anonymous, their use of so-called "home test-kits" which mean foregoing the vital interpretation of test results provided by experts in the course of genetic counseling (see also Morris 2010: 257–262).

It comes as no surprise that public opinion in different countries manifests a strong desire to keep genetic information private and to restrict its use by insurers and employers. In a poll conducted in the United States in 2006, 65 percent of respondents expressed concern about health insurance companies, and 57 percent were troubled by the possibility of banks or financial institutions gaining unauthorized access to genetic information. Furthermore, 85 percent were convinced that without laws to protect individual genetic privacy employers would discriminate. This poll documents not only a considerable amount of mistrust concerning a large variety of organizations in using personal genetic information, but also a high level of ignorance concerning legal protection against genetic discrimination. Seventy percent of respondents stated that they did not know of laws safeguarding the privacy of genetic information (White 2006). In a web-administered survey addressed to members of the National Society of Genetic Counselors in the United States in 2008, more than 70 percent of all respondents (n=312) concluded that they would definitely or possibly consider direct-to-consumer testing for patients who were concerned about genetic discrimination or wanted to be tested anonymously (Hock et al. 2011). In a survey and a follow-up case study in the Republic of Ireland conducted in 2007, 73.4 percent of respondents (n=180) thought that life insurance companies should not be allowed to use genetic test results in deciding whether or not they would insure an applicant (Morris 2010).

In order to respond to the growing fears among the population, companies in different countries have entered into "voluntary self-regulation"; this involves defining the principles and limits concerning their use of genetic information or agreeing to a moratorium on the use of genetic test results in entrepreneurial decisions. In some countries, among them the United Kingdom, national insurance associations have resolved neither to request information procured from genetic tests up to a certain insurance sum nor to consider it when formulating insurance contracts (Wilkinson 2010; Morrison 2005: 877).[33] Even before GINA came into force, a number of companies in the United States made binding commitments to limit or refrain completely from using genetic data for entrepreneurial decisions. Thus, IBM has committed itself not to employ genetic tests for assessing aptitude or health status (Lohr 2005). In the same vein, the insurance company Aetna has decided not to use genetic information to calculate premiums or reduce coverage (Mitchell 2005; Billings 2005). However, regulations introduced by a company cannot replace legislation because they can be revoked or withdrawn at any time due to their voluntary nature, and the moratoria of the insurance industry will expire after a certain period of time.

To conclude, we can say that empirical studies and legal efforts at regulation have sensitized a wider public to the problem of genetic discrimination. The subject is currently encountering considerable interest on the part of the media, and there is a growing awareness among the population of the drawbacks of the growing genetic knowledge. However, it has also become evident that the notion of genetic discrimination, as it is used in scholarly investigations and legal texts, gives rise to a variety of conceptual, normative, and practical difficulties. These will be discussed further in the next chapter.

3 Empirical Deficits and Normative Contradictions

Problems in the Analysis of Genetic Discrimination

This chapter looks at four sets of problems that reveal significant omissions and weaknesses in the debate about genetic discrimination: (1) a lack of terminological precision, which can be detected in the use of the concept of discrimination in research practice; (2) empirical shortcomings that mean central areas of genetic discrimination are not covered; (3) conceptual problems in defining the exceptional scientific status of genetic information; and (4) normative contradictions, because the idea that genetic data deserve a degree of protection not extended to non-genetic information leads to the unjustified differential treatment of those affected.

These four sets of problems will be discussed in the following sections. The central argument I shall advance is that the scientific critique and the regulatory practices addressing genetic discrimination frequently rest on the implicit notion that genes fundamentally influence human existence and form the core of our respective personalities. This critique covertly relies on an essentialist concept of a genetic program that is deemed responsible for individual development and personal identity.

DEFINITIONAL DISCREPANCIES: THE SCOPE AND CONTENT OF GENETIC DISCRIMINATION

As we have seen, "genetic discrimination" is used in scholarly writing to signify the adverse treatment of individuals based on presumed or real genetic anomalies and is rigorously differentiated from discrimination against disabled or (chronically) ill individuals (Billings et al. 1992: 477; Natowicz, Alper, and Alper 1992: 466). All the relevant empirical studies rely on this definition of genetic discrimination, but its use in the research literature has varied greatly. This makes it very difficult to compare empirical findings.

A very broad concept of discrimination is used in the work of Lapham, Kozma, and Weiss (1996). It includes not only pre-symptomatic individuals, but also cases in which the persons in question have already fallen ill. As a result, phenotypical and genotypical characteristics are both used as the basis for defining the presence of genetic discrimination.[1] Most other studies only speak of genetic discrimination against individuals if the symptoms of the illness are very mild or do not constitute a disability or result in restrictions in performance (see Billings et al. 1992; Geller et al. 1996). By contrast, Mark A. Hall and Stephen S. Rich utilize an extremely narrow concept of discrimination in their study, as they only take into account cases in which individuals were completely asymptomatic, in other words, in which the illness was not yet manifest in any way. The different concepts of genetic discrimination lead to contrary findings: whereas Hall and Rich had difficulty documenting any cases of genetic discrimination, the study by Lapham and her colleagues states that almost half of the respondents had experienced genetic discrimination (Hall and Rich 2000: 294; Hellman 2003: 86).[2]

Not just the scope, but also the substance of the concept of genetic discrimination is a bone of scholarly contention. Although people can be treated differentially in various ways, including positive or favorable ones, genetic discrimination is regularly construed as the "unjustified differential treatment of persons owing to their genetic characteristics" (Deutscher Bundestag 2002: 288; Rothstein and Anderlik 2001). The implicit assumption guiding the literature is that adverse treatment of asymptomatic individuals on the basis of their (actual or assumed) genotype is socially unacceptable, legally unjust, or morally objectionable.

However, in addition to this majority view, there are some positions that consider genetic discrimination a legitimate form of risk differentiation.[3] An extreme position in this regard is taken in the libertarian argument advanced by Colin S. Diver and Jane Maslow Cohen (2001). They contend that the differential treatment of individuals owing to genetic factors is not only morally unproblematic and legally permissible, but also socially necessary as it allows for an economic allocation of resources and a sound calculation of risks. From this standpoint, government prohibitions on genetic discrimination prevent efficient market regulation that requires a "regime of genetic transparency" (ibid.: 1482). In the case of employment relationships, for example, the use of genetic knowledge could lead to employees who are unproductive or prone to illness not being employed in the first place and thus to cost savings:

> Hiring or promoting an under-qualified or under-productive worker is inevitably costly to the employer . . . Monitoring and corrective action require investment in supervision, and often require changes in production design or scheduling. In the meantime, the under-performing worker inflicts on the organization both demoralization costs and the opportunity costs of foregone output. For these reasons, it is almost always in the employer's interest to establish better ex ante screening mechanisms so as to select workers who will require less supervision and corrective action. (Diver and Cohen 2001: 1461)

Needless to say, the definitional discrepancies are not only of interest at the conceptual level, but are of great significance for legislative practice and social policy as well. Is genetic discrimination a rational means of risk differentiation in order to generate economic growth and prosperity, or a social evil that calls for corresponding government measures to protect individuals from it? And if the latter is the case, what individuals and groups with what diseases or disease risks should be protected against discrimination, and in what way? How broad or narrow should be the circle of those who are to benefit from legal regulations? Does having a genetic trait correlate with depression or Alzheimer's disease suffice (cf. Rothstein 2001)? Will such protective measures soon take into account genetic dispositions for characteristics or traits such as intelligence or learning disabilities? This is the point where the issue of genetic discrimination intersects with enhancement strategies and the ensuing problems of social justice (Buchanan et al. 2000; Wenz 2005).

EMPIRICAL DEFICITS: BEYOND ORGANIZATIONAL DISCRIMINATION

The studies published to date that deal with the problem of genetic discrimination have a clear emphasis. They center on organizational actors disadvantaging individuals and their families. This analytical focus means that important areas of genetic discrimination do not get equal attention. I shall scrutinize two empirical deficits more closely here. The emphasis on organizational actors leads to the neglect of interactional and indirect forms of discrimination.

Interactional Discrimination

Up until now, studies of genetic discrimination have mainly concentrated on organizational actors, on the one hand, and individuals, on the other. This juxtaposition is strongly entrenched in the juridical frame of perpetrator/victim. Literature on genetic discrimination tends to focus on organizations that refuse to sign contracts or deny qualifications to individuals. Although this is certainly an important approach, there are several limitations that result from this guiding idea.

First, it might be questioned if the focus on individuals is adequate if genetic knowledge pointing to patterns of inheritance and health risks possibly shared by family members is involved. In this light, Treloar et al. (2004) suggest including the concept of "familial discrimination" in the scope of genetic discrimination. According to these authors, such an extension would make it possible to involve untested family members or those who may not even be at direct risk of a genetic condition but who may nevertheless be confronted with prejudicial attitudes or discriminatory practices.

Second, in focusing on organizational actors such as insurers, employers, and adoption agencies, previous studies of genetic discrimination failed to take into account the dimension of personal interaction. An essential arena of discrimination is thus excluded. The analytical differentiation proposed here aims to correct this empirical deficit and to distinguish in principle between two levels of genetic discrimination: one whose roots lie in personal interaction, and another based on structural relationships.[4] The first level concerns more or less spontaneous stigmatizing or disrespectful (verbal) actions by actors (individual or collective) against other persons; at the second level, organizational structures are more significant. The basis of interactional discrimination is formed by intentional forms of behavior, but also by stereotypes, prejudicial structures, and patterns of interpretation that can lead to discriminatory actions without a conscious discriminatory purpose. Structural discrimination, in contrast, is based on well-functioning, permanently established, often formalized, and explicitly regulated organizational practices. It results from the "normal performance" of established social structures.

Even though it is important to expose discriminating practices by insurance companies, employers, adoption agencies, and other organizations, this disregards a decisive arena of genetic discrimination: disrespect and stigmatization by family, friends, and fellow human beings. This interactional discrimination has been systematically ignored in most studies to date, which means that a key field of genetic discrimination is excluded. A rare exception is the study conducted by Yvonne Bombard and her colleagues to analyze the concerns and experiences of genetic discrimination among individuals and families living with the risk of Huntington's disease. This study is based on semi-structured interviews and uses a grounded theory method. Participants not only reported experiences of differential treatment in insurance, health care, and employment, areas that have traditionally received most of the attention in empirical studies on genetic discrimination, but also described discriminatory practices in the family and the social domain. Experiences included disrespect in the family, the breaking up of romantic relationships and social 'shunning.' Hence, the authors conclude,

> [T]his report highlights the fact that individuals link disrupted interactions and patterns of behavior within the family with discrimination. Although discriminatory experiences have not been typically linked to family interactions, this form of GD [genetic discrimination] needs to be considered along with more obvious forms of GD related to insurance and employment. (2008: 287)[5]

However, placing the problem of genetic discrimination in a juridical framework has an obvious benefit. When genetic discrimination is restricted to concrete individual cases and rare exceptions, the counter-strategy and its addressees are relatively easy to determine. Legal regulations have to be drawn up to ensure that the individuals affected are protected against discriminatory practices. The use and assessment of genetic information by organizational actors has likewise to be regulated, and legitimate uses distinguished from illegitimate ones. If, on the contrary, the emphasis were to be placed on the "normality" and commonness of genetic discrimination, efforts would have to be directed at public attitudes and institutional reforms which encompass altered social practices with regard to illness and disability. Also, such an approach would have to break with the cultural dominance of genetic explanations of diseases and disease causation.

Indirect Discrimination

Although it is certainly helpful to go beyond the focus on organizational forms of exclusion, disrespect, and disadvantaging to include interactional forms of discrimination in the analysis, we have to acknowledge the limits of a person-centered and case-oriented concept of discrimination. Beyond organizational and interactional discrimination, it is necessary to introduce a third category which is relevant in this context: indirect discrimination. This should encompass all factors that indirectly shape and constrain individuals' and families' scope for decision making and their options for action. Although direct genetic discrimination remains a matter of singular cases and describes the way in which certain individuals with genetic characteristics are (mis-)treated, indirect discrimination refers to social judgments of unworthiness, structures of prejudice, and forms of disrespect addressed to all members of society.

This distinction also formed the basis of the definition proposed by the German commission of enquiry on *Law and Ethics of Modern Medicine*:

> Direct discrimination means a morally unjustifiable differential treatment or exclusion of individuals by other individuals or institutions. This would include discrimination against employees, against the insured, or against people with disabilities on the basis of genetic screening. Indirect discrimination refers to social values and norms, which express the low regard for certain individuals. This would encompass the establishment of social norms, such as for example "life value attributions" (Lebenswertzuschreibungen) based on chronic illness or disability. (Deutscher Bundestag 2002: 57)[6]

The concept of indirect discrimination[7] makes it possible to analyze which new obligations regarding decision making and which hitherto unknown moral problems arise in the context of the expansion and dissemination of genetic knowledge. How do normative expectations and notions of "genetic responsibility" restrict individuals' scope of choice and action? Will the prudent management of genetic risks become an essential component of individual health care and collective prevention policies? Which forms of genetic anomalies will in contrast be represented as "avoidable" or "inferior?" All these issues have until now remained outside the scholarly debate on genetic discrimination.[8]

A comprehensive analysis of genetic discrimination must also focus on strategies with which the persons affected anticipate negative categorization by their social environment and adapt their behavior accordingly. The most effective means of pre-empting genetic discrimination entails not drawing genetic risks to the attention of other people, let alone organizational actors such as insurance companies or employers. This "information control" is exercised in private relations as well as in professional life. The persons affected experience the act of concealing their own risk of falling ill and the related fears as a constraint, as restricting forms of communication, forcing them to keep important information about themselves and their own future secret from others (see Konrad 2003; Klitzman 2010).

Moreover, the expression of unworthiness and disrespect in practices of genetic discrimination also impacts on decisions on reproduction that go far beyond the realm of those who are directly affected by the risk of genetic disease. Although studies on genetic discrimination have to date concentrated on postnatal genetic tests and the disadvantaging of persons already born, it is worth asking whether the analysis should not also cover the field of prenatal diagnostics (selective abortion on the basis of genetic knowledge) and pre-implantation diagnostics (deliberate choice of genetically "desired" embryos) (see Neuer-Miebach 2001: 56–59; Volz 2003).[9]

This limitation is linked to another shortcoming: studies on genetic discrimination tend to avoid investigating to what extent even formally voluntary practices and symmetrical decision making may have a discriminatory effect. A core element of the new genetic knowledge consists precisely in transforming biological constrains into new individual options—options that may engender new (social) constraints. Scholars have discerned a transition from a state-led and often compulsory eugenics to more indirect forms of control and guidance of individuals, something the academic debate has termed "voluntary eugenics" (Weß 1992), "individual eugenics" (Waldschmidt 1996: 275), or "liberal eugenics" (Habermas 2003). These accounts point to the fact that individual decisions and private choices may, taken collectively, have undesired social effects and create new pressures on reproductive behavior. It is certainly debatable whether or not contemporary human genetics and reproductive medicine represent a disguised or a new kind of eugenics (Paul 1998a; Lemke 2002; Koch 2004; Rose 2007). However, it should be obvious that genetic discrimination is a complex and multi-faceted problem that cannot be reduced to practices of prohibition and exclusion alone. Instead of being confined to single cases and rare exceptions, it may well result from "normal" risk assessments, "rational" concepts of health and a "self-regulative behavioral management" (Weingart 1992: 45–49).

There is another aspect that has to be taken into account. It is not enough for a systematic analysis of genetic discrimination to simply internally differentiate between various forms and fields of genetic discrimination. It is also necessary to consider the "structural coupling"[10] between different forms of discrimination. In this way, it is possible to study how forms of genetic discrimination link up with sexist and racist practices and mutually reinforce one another (Wolf 1995; Andrews 2001: 77–97).[11]

As regards the relationship of genetic discrimination and racism, it bears stating that some genetic diseases are more prevalent in certain ethnic groups than in others. For example, sickle cell anemia occurs more frequently among individuals of African descent, Tay-Sachs Syndrome is especially widespread among Ashkenazi Jews, and most patients living with the beta thalassemia gene are inhabitants of the Mediterranean rim. Because certain ethnic groups are differently susceptible to particular genetic illnesses, there is a danger that members of minorities will be associated with such genes and treated as if they were ill, even if they do not bear the particular genetic mutation. For example, sickle cell anemia in the United States is considered a disease of African-Americans, although it is to be found just as frequently among population groups from the Mediterranean rim (Duster 1990: 24–28, 45–51). There is also a danger that there will be a disparity in resource allocation and in the public attention paid to these diseases. For example, in the United States research into cystic fibrosis, which mainly affects white people, receives far more financial backing than research into sickle cell anemia, even if the latter is far more widespread among the general population (European Commission 2004: 50).[12]

The practice of "gender verification" at international sports competitions is a striking example of the linking of genetic discrimination and sexism. This procedure has been used since the 1960s to determine the sex of female athletes. It was first introduced by the International Amateur Athletic Foundation (IAAF) in 1966 at the European Championships in Track and Field, when rumors had been circulating that men were masquerading as women in women's events. From 1968 onward, "gender verification" was obligatory for women participating in the Olympic Games (since 2000, gender testing has not been systematically used, but is only applied in "cases of doubt"), and it is still conducted in some disciplines in international competitions to "check" the biological sex. This control mechanism resulted in a series of women with seemingly "male attributes" (e.g. females with androgynous insensitivity, XY gonadal dysgenesis) who were excluded from the competition. Only women and not men have to have their "gender" verified, and suffer the possible consequences of "failing" such a test. In most cases, the women affected only learned from the test procedure itself of their "anomaly" (Simpson et al. 2000; Dickinson et al. 2002; Müller 2006; Heggie 2010; Hercher 2010; see also Lemke 2004a).

In sum, a person-centered and case-oriented concept of discrimination is not able to explore the systematic links between racist and sexist practices and ideologies, on the one hand, and geneticized patterns of interpretation and modes of action, on the other. Genetic discrimination does not entail isolated and accidental deviations from the norm, individual cases, and erroneous organizational processes, but it points to social practices that divide persons into genetic categories and promote a belief in the determining power of genes (Wolf 1995). Unfortunately, notions of genetic essentialism frequently also inform the analysis and critique of genetic discrimination, leading to conceptual confusion on the one hand and normative ambivalences on the other. These problems will be addressed in the next sections of this chapter.

CONCEPTUAL CONFUSIONS: THE EXCEPTIONAL EPISTEMOLOGICAL STATUS OF GENETIC INFORMATION

Empirical studies on genetic discrimination and legislative initiatives to protect those at risk of being discriminated against due to their genetic propensities often share an important assumption: both take for granted the exceptional status of genetic information. The idea of "genetic exceptionalism" entails two different dimensions.[13] First, there is the claim that genetic information differs epistemologically from non-genetic information. From this perspective, there are clear scientific criteria separating genetic testing from non-genetic diagnostic or predictive procedures, genetic diseases from non-genetic conditions. Second, there is the suggestion that genetic information should be normatively distinguished from non-genetic information. According to this view, discrimination on the basis of genetic data on the future health of an individual is (more) unfair compared to discrimination based on non-genetic medical information (Alper and Beckwith 1998; Hodge 2004).

As I shall attempt to show, both assumptions provide an oversimplified account of the complex issues at stake and the two distinctions appear more blurred. I shall focus in this section on the justifications offered in support of the idea of an exceptional epistemological status for genetic information, and the next section will trace the normative problems to which this perspective gives rise.

Genetic Exceptionalism

There are at least three arguments put forward to justify the exceptional (medical) position of genetic testing. It is claimed that genetic tests are more precise and reliable than other medical tests in providing predictive information on the health of an individual. They are said to determine with certainty or high probability whether a person will fall ill with a specific disease. Second, the results of genetic tests are said to enable conclusions to be drawn about the state of health or disease risks of family members related to the individual who is tested. Genetic tests are, third, believed to differ from traditional diagnostic techniques and conventional medical procedures as they purportedly reveal fundamental and immutable personal characteristics of the individual examined. However, none of these three criteria used to justify "genetic exceptionalism" stand up to closer scrutiny (Murray 1997; Lemmens 2000; Kollek and Lemke 2008: 53–71).

First, only a very few genetic tests allow conclusive predictive statements to be made about future illnesses. Genetic diseases are as a rule characterized by incomplete penetration or variable expressivity (see on the following Alper and Beckwith 1998: 143–144). The former relates to the frequency with which a genetic mutation actually triggers the corresponding disease in the person carrying the gene. In the case of complex illnesses such as cancer, Alzheimer's, or diabetes, the presence of a modified gene does not necessarily lead to illness. For example, proof of a mutation of one of the so-called BRCA genes increases the statistical probability of a woman developing breast cancer, but the question of whether, when, and in what way she will be affected by the illness is by no means answered (Feuerstein and Kollek: 2000; Lemke 2004b: 70–76). As studies have shown, even monogenetic illnesses such as cystic fibrosis or Huntington's disease do not entail 100 percent penetration (see Kerr 2000; Van den Boer-van den Berg and Maat-Kievit 2001; Thauvin-Robinet et al. 2009; Sequeiros et al. 2010).

Variable expressivity refers to the fact that the symptoms of one and the same disorder can differ greatly from one individual to the next, although both possess the same modified gene. In other words, the same DNA mutation can trigger quite different clinical symptoms—or none at all; conversely, one and the same disease can be triggered by different genetic variations (Hubbard and Lewontin 1996; Wolf 1997). It bears stating here not only that the predictive value of genetic information is often exaggerated in public discourse and sometimes even in scientific debates, but also that non-genetic medical tests provide information on future health risks, such as an HIV test, a cholesterol screen, or the proof of an asymptomatic hepatitis B infection. Likewise, blood-pressure tests or proof of blood in stool are of great diagnostic importance when identifying illnesses at an early stage before the first symptoms of coronary heart disease or stomach cancer have appeared (Murray 1997: 64; Zimmern 1999; European Commission 2004: 44; Kollek and Lemke 2008: 53–71).

As regards the second criterion, it is also difficult to specify the medical uniqueness or even any special status of genetic analyses. As Murray rightly emphasizes, people have long since known that their own risk of having an illness increases if the same illness has already arisen in their family. Likewise, in the past, doctors have relied on information on relatives to supplement their medical knowledge about a particular patient. Moreover, an individual's family history has for some time been factored into insurance contracts or consulted when defining premiums. Genetic tests are, in other words, neither the only nor the most important tools for using medical data about relatives in order to reach statements on an individual's health status (Murray 1997: 65).

The third criterion rests on the idea that genes are autonomous and active, forming a kind of control center that steers and regulates the organism. This assumption is at best a considerable simplification, if not misleading or false. Genes are not static units, but an integral part of a complicated biochemical network that is defined by the dynamic interaction of interdependent actors (see Keller 2000; 2002; Moss 2003). For this reason, it is an inadequate simplification if we assume that a characteristic or function is determined in part by heredity and otherwise by environmental influences. This fails to consider the possibility that genetic changes can also be first *acquired* in the course of life owing to environmental influences and a specific lifestyle.[14] Furthermore, there can be no simple equation between genetic causality and an inevitable fate or non-genetic factors and individual choice. Genetic causes of illness do not signal necessity and biological determination, and non-genetic conditions are not in principle easier to control or the result of personal choices for which the individual is responsible.[15] On the contrary, there are also many non-genetic factors which individuals can by no means control: "If the air we breathe and the water we drink are polluted, if our parents or co-workers are heavy smokers, if we are reasonably prudent but injured in an accident nonetheless, it is hard to say that we bear any significant measure of responsibility for the resulting illness." (Murray 1997: 65–66; cf. Geetter 2002)[16]

All three criteria that ostensibly justify the privileged position of genetic information discussed here thus prove unconvincing. As no scientific basis for the exceptional nature of genetic information exists, this status must rest not on an intrinsic quality but solely on its social and cultural significance.

> Genetic information is special because we are inclined to treat it as mysterious, as having exceptional potency or significance, not because it differs in some fundamental way from all other sorts of information about us . . . I propose that genetic exceptionalism—the plea to treat genetic information as different from other health-related information—is an overtly dramatic view of the significance of genetic information in our lives. It is a reflection of genetic determinism and genetic reductionism at least as much as the product of genuinely distinctive features of genetic information. (Murray 1997: 71)

That said, not only is it hard to be sure where the line dividing genetic from non-genetic factors runs, but it is also not clear what exactly "genetic" means. In other words, what we have here is not a definitional problem that results from an inacceptable conflation or erroneous shift of the line dividing two distinguishable components (genetic/non-genetic); instead, it is a systematic problem that lies in the "nature" of the matter at hand. Genes are not naturally defined units whose empirical characteristics are fixed and must merely be made visible by scientific methods. They are not simple biological facts, but rather "epistemic things" (Rheinberger 1997a). Genes are the product of the interaction between material practices and cultural patterns of interpretation, of entanglements between matter and meaning; they are not the starting point, but the result of a co-production of material arrangements and discursive practices (Kollek 2002; Moss 2003; Barad 2007). What is respectively termed "gene" or "genetic" is marked by great semantic flexibility and depends on scientific definitions and social contexts (Kay 2000). This can be demonstrated with the help of two examples. As discussed above (in Chapter 1), it is not possible to conclusively and bindingly define what distinguishes a genetic disease from a non-genetic disease; moreover, it is not clear how genetic tests can be accurately differentiated from non-genetic ones.

Genetic or Non-Genetic?

As we saw in Chapter 1, the concept of genetic disease has been expanded in recent decades. Today, genetic factors are considered to be responsible not only for "monogenetic" diseases but also for multi-factored conditions such as cancer, Alzheimer's, diabetes, and many other common diseases. This semantic expansion has gone hand in hand with a shift in meaning such that diseases are increasingly construed as genetic deviations from the norm. However, if every disease can potentially be ascribed to genetic variations, it becomes questionable what exactly the contribution made by genetic factors is and how genetic causes of diseases are to be distinguished from their non-genetic counterparts (Yoxen 1984; Lemke 2003).

But not only is it impossible to finally and definitively state what distinguishes a genetic illness from a non-genetic condition; it is equally unclear how genetic tests can be distinguished sharply from non-genetic analyses. "Genetic diagnostics" is the umbrella term used to cover all types of testing for genetic diseases or altered forms of genes associated with an increased risk of developing genetic disorders. This includes tests at the level of DNA as well as phenotypical diagnoses (such as the test for genetically conditioned color blindness), chromosome examinations, and analyses at the level of gene products (Zimmern 1999; Schmidtke and Sperling 2003; Sequeiros 2012; Varga 2012).

Joseph S. Alper and Jon Beckwith rightly point out that this comprehensive definition of genetic tests applies to practically all clinical examinations. They emphasize that most medical tests focus on diagnosing abnormal concentrations of biochemical units. Typically, these units are proteins that are generated directly by genes, or molecules whose synthesis depends to a more or less large degree on the activity of the genes. This is why a specific biochemical finding can be the result of a changed gene or genotype. Even if these non-genetic tests focus on determining the function of the various organs, they may also provide information on genetic characteristics. In addition, in some cases only the medical context can decide whether a genetic or a non-genetic test is involved: "A test of blood cholesterol concentration may be regarded as genetic when testing an individual with a family history of hypercholesterolemia, a single gene recessive disorder. The same test is clearly a non-genetic medical test when ordered in the course of a routine physical examination." (Alper and Beckwith 1998: 145; Beckwith and Alper 1998: 207)[17]

It follows that (mono-)genetic diseases cannot be exclusively detected by means of genetic testing. In many cases, conventional screening procedures can reliably prove the existence of a genetic condition. That the distinction between genetic and other diagnostic devices is not merely a technical matter can be seen from the history of the phenylketonuria (PKU) screening as reconstructed by Diane Paul. As we saw in Chapter 1, PKU is a metabolic illness that can be prevented by initiating a special diet when the patient is still a child. When, in the 1960s, the so-called Guthrie test was developed and used as a diagnostic tool with newborn infants, the genetic dimension of the disease played no role; instead, back then PKU was considered a treatable form of mental retardation. Moreover, the diagnostic method applied is a biochemical test, not a genetic analysis. Only much later did the Guthrie test come to be regarded as a genetic test and PKU as a genetic condition—a label that served various interests. The molecular medicine visionaries were able to point to the PKU test to demonstrate that genetic tests with a therapeutic benefit existed. Critics of genetic screening also profited from the label "genetic test." In the framework of the Human Genome Project, commissions were set up in the United States and various other countries to specifically regulate genetic analyses (in contrast to other medical tests). Because genetic screening requires greater technical care and more intensive legal regulation than other methods, the definition of the PKU test as "genetic" was also of interest to critics of genetic determinism who were skeptical of the increasing use of genetic screening (Paul 1999; on the history of PKU, see Paul 1998c).

It is thus difficult to distinguish with any scientific precision between genetic and non-genetic diseases, and the dividing line between genetic and non-genetic tests cannot be defined intrinsically in technical terms either. As we have seen, in the final instance the difference is the product of social debates and scientific compromises. This insight has a strong impact on political initiatives that endeavor to regulate the use of genetic data by means of special non-discrimination legislation.

NORMATIVE AMBIVALENCES: THE LEGAL PRIVILEGE OF GENETIC INFORMATION

Discrimination on the basis of genetic factors is directed against "asymptomatic ill" (Billings et al. 1992), whereas discrimination against chronically ill and disabled individuals results in adverse treatment on the basis of phenotypic features. Generally, the former is considered more problematic in moral and legal terms than the latter, which leads to asymmetrical treatment of those suffering from discriminatory practices. First, genetic discrimination and discrimination based on phenotype or on already manifest symptoms are assessed differently in legal terms, which prompts the question of what criteria are used to justify such differential treatment of persons who are equally affected by discriminatory practices. Second, there is the danger that the exceptional legal status of genetic discrimination simply "normalizes" all non-genetic forms of discrimination. If in the widest variety of different social areas the disabled and the (chronically) ill are regularly discriminated against compared with the healthy, then this appears legitimate to the extent that special protection exists for persons who are affected by practices of genetic discrimination.[18] In other words, the focus on the legal impermissibility and moral reprehensibility of genetic discrimination isolates other forms of discrimination, plays them off against one another, and threatens to augment the social acceptance of practices of non-genetic discrimination.[19]

Gregor Wolbring (2001: 87–88) has pointed out that in principle, two completely different strategies for legal policy are conceivable as ways of preventing discrimination against the asymptomatically ill. One path could be to extend and expand existing non-discrimination legislation on the equal treatment of the disabled in such a way that it also covers discrimination against the asymptomatic and symptomatic ill, individuals affected by genetic and those affected by non-genetic diseases. However, at present a quite different approach is being followed in the United States and many other countries, one that favors special laws to protect the asymptomatic ill based on genetic characteristics by stressing a strict and unequivocal legal distinction between asymptomatic and symptomatic individuals. Wolbring points out that this policy serves to increase the legal gap between the two groups. Genetic discrimination is not treated as an integral part of a social continuum of discriminatory practices, but is considered as a specific caesura to be distinguished in conceptual and normative terms from other forms of disadvantaging. As Wolbring rightly remarks, it is questionable whether discrimination against the asymptomatic ill based on genetic characteristics would exist without discrimination against the symptomatic ill and the disabled (Wolbring 2001: 87).

Strikingly, there are differences in the legal and moral appraisal not only as regards the symptomatic as opposed to the asymptomatic ill, but also within the group of the asymptomatic ill. A distinction is made there between individuals affected by genetic risks and those exposed to non-genetic disease risks (e.g. lifestyle risks like smoking). In both groups the illness has not yet become manifest and will possibly never become so, yet the legal assessment of the risks is different. The problem is perhaps best illustrated by juxtaposing two similar cases that were decided almost simultaneously in the same federal state of Germany.

"From the Cradle"—the Rebirth of Genetic Essentialism as Non-Discrimination Policy

The first case was taken up by various media and attracted a lot of attention inside and outside Germany (Mechan-Schmidt 2003; Burgermeister 2003; Traufetter 2003). In August 2003, the State of Hesse refused to employ a teacher as a civil servant after she had completed her trial period. The enquiry by the officially appointed occupational physician had revealed that the young woman's father lived with Huntington's disease. The report came to the conclusion that at the present point in time the applicant's health was suitable to enable her to take up the job, but she was barred from becoming a civil servant on the ground that there was an increased probability that she would fall ill in the foreseeable future and become permanently unfit to discharge her duties. The applicant lodged an appeal against this decision before the Administrative Court in Darmstadt, which ruled mainly in her favor and instructed the State of Hesse to immediately appoint her to government service. In the court's opinion, the school authorities had wrongly assessed the state of her health as an applicant, as they had claimed that the 50 percent risk of illness meant there was a "most strong probability" she would not be able to discharge her duties. The school board declined to contest the decision and agreed to employ the woman in question as a civil servant (Frankfurter Rundschau 2004a).[20]

At the same time, another case was decided in court that attracted far less public attention. Again in Hesse, a young man was dismissed while still on probation for government service, as in the opinion of his employers his weight of 120 kg. meant that his health was not suited for a career as a civil servant in public administration. Here again the applicant took the matter to court. However, in its ruling the Frankfurt Administrative Court confirmed that the dismissal was legal as the employer was permitted to pre-empt the risk of having to foot the bill for later enduring damage to the man's health (Frankfurter Rundschau 2004b). In other words, although the applicant was not yet ill, and it is completely uncertain whether and in what way his body weight might impair his health in the future, he was treated in legal terms as though he were already suffering from a pathological condition and incapable of carrying out his duties.[21]

These examples show that the distinction between genetic[22] and non-genetic information can hardly serve as the basis for legal differentiation and may have problematic social implications. Several arguments have to be taken into account. First, it is hard to see why, for example, the use of biochemical methods that permit inferences about a person's genetic disposition is permissible for discriminatory purposes, whereas genetic tests that reach the same results are forbidden. It would seem not only impractical but also unfair to prohibit an insurance company from considering a genetic analysis for a complex disease when the results of a non-genetic test for the same disease may be relied on. This approach enacts a legal double standard in which people with positive genetic diagnoses receive more protection against discrimination and data abuse than those whose diagnoses are based on non-genetic methods:

> Would the law mean that the records of a person with a presymptomatic heart condition who was given a genetic test for some mutant gene associated with heart disease would be covered by anti-discrimination provisions, but not the records of a person with the same condition whose physician orders only nongenetic tests? (Alper and Beckwith 1998: 147)[23]

Second, it remains unclear why organizational actors such as insurance companies or employers should be forbidden from using *one* source of genetic information (genetic testing) although they are allowed to draw on other forms of genetic knowledge. For example, according to the German Genetic Diagnostics Act an insurance company would be prohibited from demanding that a woman who a BRCA test shows as positive for breast cancer pay a higher insurance premium, whereas this is permissible with regard to a woman who has preferred not to undergo genetic testing but in whose family several women have already had breast cancer (Backhaus 2012; see also Geetter 2002). This asymmetry in considering relevant parameters in risk assessment not only violates the principle of fairness, but might also compel people to opt for genetic analyses even if they did not originally want to—for example, in order to get insurance policies (at standard conditions).[24]

Third, the ongoing discovery of genetic factors for the development of diseases will in future make it ever harder to draw a line between genetic and non-genetic conditions. This is particularly evident in the genetics of mental disorders. The search for candidate genes for major psychiatric disorders like schizophrenia or depression is complemented by the increasing acknowledgement of complex interactional patterns involving genetic factors, psychosocial influences, and environmental exposures in accounting for the manifestation of a particular disorder (Appelbaum 2010). The same development could be detected with regard to so-called lifestyle diseases. With reference to the case mentioned above, a series of research findings could be cited, for example, that point to a genetic component in obesity (see Shell 2002; Wardle et al. 2008). What would the judgment be if an applicant could credibly claim that a specific genetic disposition is (co-)

responsible for his increased body weight? Would we then be confronted by a case of genetic discrimination, and would we need to provide special protection to those affected by it?

To summarize, we can say that the emphasis of non-discrimination and privacy policies is evidently more on the (genetic) "nature" of information and less on social practices in which these data are taken up and used (see also Wolf 1995). Instead of reproducing and stressing an essentially artificial distinction between genetic and non-genetic information, policies should consider the consequences of the dissemination, use and storage of all types of medical (predictive) information. Legislation to protect against genetic discrimination rests often on the idea that genes fundamentally influence human existence and constitute the core of each personality. It is therefore seen as unfair to punish persons for something they cannot themselves control.[25] This logic is also to be found in the justifications given by the German Health Ministry for enacting a particular law on genetic diagnostics. The statement in question declares that any form of genetic discrimination must be rejected for the following reason: "Our genetic characteristics 'are with us in the cradle' and we are therefore not responsible for them." (Bundesministerium für Gesundheit und Soziales 2004: 16)[26] This argument may initially seem plausible, but on closer inspection it is not very convincing as it is based on the idea of a genetic program that is "responsible" for the development and identity of individuals and that both constitutes and constrains their scope for action. The assumption of a particular power and autonomy of genes prepares the ground for their privileged legal status. In principle, there are many non-genetic factors that are just as little susceptible to personal control although we do not call for them to be specifically protected. It is by no means evident why a person whose higher risk of being affected by a specific type of cancer, say, is attributable to genetic factors, should enjoy greater protection than someone whose health is threatened by environmental factors such as poor working conditions or polluted air. Should not this group of persons likewise be effectively protected (see Diver and Cohen 2001: 1451–1452)?

The hypothesis that genetic discrimination is apparently less tolerated than other practices of social discrimination and disrespect is corroborated by the results of a study conducted by Timur Kuran and Edward J. McCaffery. In 2002, these authors interviewed 1,500 U.S. citizens by telephone on their attitudes to different forms of discrimination (this was supplemented by an equally large group that answered an online questionnaire). The questions addressed the acceptance of five types of discrimination: travel limitations for U.S. air passengers of Arabic origin; the possibilities of employers refusing to hire overweight persons; the genetic screening of job applicants for possible medical risks by employers; stricter police controls for car drivers of Afro-American origin; and the consideration of education in deciding whom to admit as an immigrant. The results not only demonstrated that in all the cases men were more willing than women to consider discrimination as legitimate, but also revealed a marked hierarchization of the social acceptance of discriminating practices. Those surveyed sharply distinguished between the distinctive forms of discrimination and showed the most understanding for discriminating against poorly educated immigrants (36.8 percent of men and 26 percent of women), whereas genetic discrimination met with the least acceptance—only 7.8 percent of men and 2.8 percent of women approved of it (Kuran and McCaffery 2005).

The explicit withdrawal of responsibility in the case of genetic disease risks might only be the other side of the coin of an increasing ascription of responsibility for all factors considered non-genetic.[27] Support for this interpretation is provided by an empirical study on the way life insurers account for predictive medicine (e.g. genetics, lifestyle) in medical underwriting (Van Hoyweghen, Horstman and Schepers 2006). The authors of this study did ethnographic fieldwork in Belgian life insurance companies in the context of the Belgian prohibition on the use of genetic data in insurance. They found that their use of predictive medical information emphasized responsible health behavior, which prompted them to focus on lifestyle risks more than on genetic risks. Although all insurance companies included in the study asked for family history in the medical questionnaire or requested this type of information from GPs or specialists, the information was rarely used in underwriting policies. Family history played a minor role in underwriting compared to lifestyle risks.

> Charging somebody because of his genes or family's health condition alone is considered un-delicate or unjustifiable . . . Where genetic risks are seen as "fate", as "involuntary" or "uncontrollable", lifestyle risks are considered as self-induced, voluntary and "own responsibility". The outcome is a financial solidarity or collective responsibility for the genetic *risk carriers*—the collective risk pool is prepared to pay for them—and individual financial responsibility for lifestyle *risk takers*—they have to pay themselves for their risk via extra premiums. (Hoyweghen, Horstman and Schepers 2006 1232–3, emphasis in original; see also van Hoyweghen 2007; Pollitz et al. 2007)

The reductionist concept of genetic fatalism as something that purportedly unravels automatically and independently of the individual person's will contrasts with the radicalized appeal to individual responsibility and personal accountability as regards health and the prevention of illness (Kühn 1998; Bause 2000; Crawshaw 2012). It may be that the different judgments in the two cases mentioned above stem from a common underlying logic that places the question of personal responsibility at the center of things. Increased body weight and the resulting risks to a person's health are essentially considered the (erroneous) result of individual choice, whereas the risk of being affected by Huntington's disease is viewed as biological fate and thus treated as something for which a person cannot be held responsible.

I investigate the relationship between the concept of responsibility and genetic risks in more detail in Chapter 5. Before that, however, the results of the first German study on experiences of genetic discrimination will be presented.

4 "A Slap in the Face"
An Exploratory Study of Genetic Discrimination in Germany

In contrast to the United States, the United Kingdom, and Australia, there are no empirical studies on genetic discrimination in Germany. For this reason, it remains unclear how often individuals and families in Germany are discriminated against because of their genetic characteristics. Aside from individual cases, do forms of systematic genetic discrimination exist? If so, in which organizational settings and social arenas do they appear? These questions cannot be answered at present even cursorily, because no empirical studies or basic surveys have been carried out. This absence of research on genetic discrimination in Germany is all the more surprising as the question of discrimination plays a key role in the public discussion of genetic diagnosis and its social impact. The problem is repeatedly highlighted in press reports and media coverage (*Der Spiegel* 2004; Götte 2004; Schwägerl 2004; Gelinsky 2008). The German National Ethics Council has dealt with the issue of genetic discrimination in two of its statements on predictive genetic testing, focusing on discrimination by employers and insurers (Nationaler Ethikrat 2005, 2007).

Although genetic discrimination gives rise to heated exchanges in political and media debates in Germany, it remains a non-issue in the academic literature. The study presented in this chapter is intended to draw attention to this gap in research, and to contribute to closing it in part. It focuses on persons who are affected by (the risk of) Huntington's disease (HD)[1] and their experiences of genetic discrimination. The presentation that follows is not a systematic investigation, however, but an explorative collection of case studies that could form the basis for a preliminary assessment of the specific domains and forms of genetic discrimination in Germany.[2]

The findings indicate that earlier studies on genetic discrimination in different countries have used the concept of genetic discrimination too narrowly, often referring only to organizational discrimination. I will argue that it is necessary to broaden the concept by including, first, mechanisms of indirect discrimination and, second, forms of interactional discrimination. Furthermore, the findings indicate that the strategies individuals employ to manage and minimize the risk of genetic discrimination have to be taken into account to get a more complete picture of genetic discrimination.

The chapter starts by presenting the research approach adopted for the study and explains why the analysis concentrated on HD. The following sections focus on the research findings and discuss their implications for the concept of genetic discrimination. The concluding section summarizes the main results and outlines some limits of the study.

METHODOLOGY AND RESEARCH APPROACH

Huntington's Disease

Several considerations were crucial for the decision to approach individuals affected by (the risk of) HD for information about their experiences with genetic discrimination. The first concerns the etiology of the disease. HD is a late onset disease that usually first manifests itself in the fourth or fifth decade of an individual's life. The condition generally entails severe physical and mental changes. The beginning of the disease is frequently marked by involuntary, sudden muscular twitching throughout the body or psychological disturbances and changes in personality; as the disease aggravates, the person's physical and mental state deteriorates. Usually the symptoms progress until death after ten to fifteen years, but longer periods of illness have been known, some extending into very old age (Roos 2010).

HD is an autosomal-dominant hereditary disease, that is, statistically every child of an individual carrying the mutation has a 50 percent probability of receiving the allele[3] in question. Because, in the vast majority of cases, HD does not manifest itself until late in life, the children of HD patients live with uncertainty as to whether or not they themselves carry the mutation and will be affected by the disease in the future (Dose 1997; Lange 2002; Shoulson and Young 2011). Although these "persons at risk"[4] are not ill and may never develop the disease, they are already exposed to discriminating practices, as studies in several countries have documented (see Chapter 2). Equally uncertain is the further course and duration of the disease. The latter is on average twelve years, but less than nine or more than sixteen years among almost half of the patients (Wexler 1992: 212; Schmidtke 1997: 45–47; Roos 2010).

In Germany, approximately 7,000–8,000 persons are directly affected by HD (the disease occurs with a frequency of 1 in 10,000 to 1 in 12,000), and the number of "persons at risk" who may fall ill at a later stage in their life is several times higher. The disease is triggered by a multiple repetition of a specific triplet, a trinucleotide repeat of the bases cytosine–adenine–guanine (CAG), on the short arm of chromosome 4. Those carrying the gene for HD usually have more than thirty-eight of these repetitions, control individuals less than thirty (The Huntington's Disease Collaborative Group 1993). The disease has almost complete penetrance. Every "carrier" of the mutated gene has a high probability of becoming ill, but it is unclear when the disease will manifest itself and how serious it will be.[5] Although the genetic locus of the disease was identified some time ago, to date it has not proved possible to develop any effective therapy. Pharmacological and physiological treatment, cognitive training processes, and supportive psychotherapeutic consultation may relieve the symptoms of the illness, but there has been little success in preventing the process of decline and the death of the patient. Most strategies still focus on the aim of delaying the onset and slowing the progression of HD (Abdulrahman 2011).

The second reason for choosing HD in this study is that a predictive genetic test has existed for two decades which enables the mutation to be detected in blood or tissue samples (The Huntington's Disease Collaborative Group 1993; Kremer et al. 1994). The genetic test for HD was long considered *the* definitive test for the prediction of a genetic disorder, because it was assumed that a negative test result meant that the risk of disease was definitively ruled out, whereas proof of the mutation indicative of HD meant that the "carrier" would inevitably fall ill. However, research findings show that at least in some cases the test results do not provide an unequivocal conclusion (cf. Van den Boer-van den Berg and Maat-Kievit 2001). The "diagnostic invitation" (Kenen 1996) gives "persons at risk" the opportunity to find out with relative certainty whether or not they carry the mutation. In view of the lack of therapeutic possibilities, the predictive testing of "persons at risk" cannot be used to avoid the illness or alleviate its symptoms, but solely for their family or life planning. At the same time, the abstract possibility of verifying the risk status may lead third parties such as insurance companies or employers to exert significant pressure on a potentially affected person to inform him or her about his or her risk.[6]

Third, there is evidence suggesting that "persons at risk" for HD have already experienced genetic discrimination in Germany. Although only a few incidents have been reported in the media, it is well known inside support groups that more cases exist. Thus, Mrs. Erika Benderoth from the local Berlin group of the Deutsche Huntington-Hilfe (DHH) [German Huntington's Disease Assistance] reported to the study commission of the German parliament "cases of bullying, raising of insurance premiums for those insured or cancellation of policies." She also referred to another problem: the misuse of personal data in connection with the predictive genetic test.

> Since our experience has shown that unfortunately there are leaks in the insurance companies, and the issues of data protection and confidentiality are sometimes not taken too seriously, persons at risk who want to take the test are forced to pay for the expensive examination from their own pockets . . . An example from the self-help group: One woman who adopted a child many years ago was recently informed

by the youth authorities that the father of this child allegedly had Huntington's chorea. (Enquete-Kommission 2000: 54)

Interestingly, among the few cases known at present practices of genetic discrimination among public employers are prominently represented. The case of a teacher who the federal state of Hesse refused to hire as a civil servant on probation in 2003 has already been mentioned briefly in Chapter 3. The young woman is apparently the only person who has legally challenged the refusal of civil servant status due to genetic disease risks. Two more cases have been reported by the DHH. At the beginning of 2000, a young policeman in Bavaria was not granted civil servant status due to his family history, as one of his parents lived with HD. Although the man was demonstrably healthy at the time of application, he was denied civil servant status on the basis of his 50 percent chance of falling ill. The justification utilizes the same reasoning that has been brought forward in other civil service cases against individuals with a family history of HD:

> The health requirements for later acceptance into the civil service status for life or for unrestricted professional activity for several decades as a police officer are not fulfilled when according to medical and scientific opinion and experience it is to be expected that the disease or disorder will not foreseeably allow *long-term* unrestricted employment in full police service before the end of the regular duration of career service.[7]

The 50 percent probability of developing HD has thus been transformed in this decision into something "expectable" and "foreseeable." Because the applicant nonetheless wanted to pursue his professional career as a police officer, he saw no other alternative than to determine through the predictive genetic test whether or not he carried the mutation. After the test result proved to be negative, he was ultimately granted civil service status (see also Zerres 2001: 246).

A similar case from 2003 exhibits the bizarre forms genetic risk calculation can take and what disturbing results it may lead to.[8] Here, too, a young man was refused admittance to the police service after a health examination, because he had truthfully stated that his grandfather lived with HD. The applicant, however, had statistically only a 25 percent risk of falling ill, because his father, with a risk probability of 50 percent, had up to that point not shown signs of HD. In other words, even a 25 percent probability of risk that the disease would appear in the future seemed to the police authorities "foreseeable" and led to them denying an applicant civil servant status. Particularly interesting in this case is the fact that the young man's father himself was a civil servant in the police service. He probably had not been asked years before about genetic risks or at that time did not know about the case of HD in his family.

Study Design

Empirical investigations on experiences of genetic discrimination confront a twofold problem. how is it possible to identify those affected, and how can these individuals be persuaded to provide information on discriminatory practices? Some studies have worked successfully with self-help groups, whose spectrum of members includes not only individuals affected by the corresponding disorder but also "persons at risk" and their family members (cf. Geller et al. 1996; Low, Kind, and Wilkie 1998). Such cooperation is certainly not without problems. It can lead to potential biases as regards the study's results because the participants may be considered a self-selected group and the problem of promoting organizational interests within the framework of a scientific study might occur (on this problem see Treloar et al. 2004: 164).

Despite these concerns, this approach was chosen for the present study. It was conducted in close cooperation with representatives from the DHH, the German HD self-help organization, which supports and coordinates the work of regional self-help groups in Germany. Two reasons were crucial for this decision. First, the DHH is less a hierarchically structured lobby organization than a coordinating office with a decentralized administrative structure connecting autonomous regional groups with one another. Second, there was no alternative to this form of cooperation as regards the goal of the study, because it would otherwise have been extremely difficult, if not impossible, to approach those affected to provide information on their experiences of genetic discrimination. The informants' trust in the responsible collection, storage, and evaluation of sensitive data is an indispensable requirement for the success of such a study.

The DHH currently has some 1,500 members and makes contact with around another 300 families every year. The group is organized into a national section, several regional sections, and numerous local self-help initiatives. A scientific council comprising researchers in the field of HD serves as an advisory committee. Its function is to keep the group abreast of the latest findings with regard to the disease (www.huntington-hilfe.de [accessed May 2, 2012]; Deutsche Huntington-Hilfe 2001: 57–59).

The concept of the study was submitted to the DHH management board and to the advisory council. Once both bodies had agreed to support the study, a questionnaire was published in the DHH quarterly *Huntington-Kurier* (No. 2, 2004). In an accompanying text, the author introduced himself and his research activity and explained the study's aims and approach. He also assured those prepared to take part that the information they provided would be used anonymously and in accordance with data protection regulations.[9] The same issue of *Huntington-Kurier* carried a joint statement by the management board and the advisory council of the DHH, in which they encouraged readers to fill in the questionnaire and to participate in the study.

In order to increase the number of respondents, questionnaires were sent to approximately thirty contact groups in Germany and Austria. These groups usually meet on a monthly basis and provide emotional support to those affected, discuss practical problems, and exchange information on social and medical questions and on the effects of the disease on family life. The questionnaires were addressed to the organizers of the contact groups with a request to hand them on to individuals willing to participate in the study.

The questionnaire combined closed and open questions.[10] The former referred to the person's general health and risk status, whereas the latter asked for a description of possible instances of genetic discrimination. The questionnaire covered a broad spectrum of issues that might be relevant for discriminatory practices. Whereas most investigations concentrate on discrimination on the part of organizations (such as insurance companies and employers), participants in this study could also make reference to discriminatory practices among their circle of friends and acquaintances, and even within their own family.

The evaluation of the returned questionnaires sought to ascertain the prevalence and the specific forms of genetic discrimination. If signs of experience of genetic discrimination were present in the questionnaire, telephone interviews were conducted with the respondents (if they had granted permission beforehand). The goal of this supplementary questioning was to gain further and more detailed information on possible discriminatory practices. Altogether the author conducted nine interviews lasting between twelve minutes and more than an hour. Because not all respondents had agreed to engage in a telephone interview, in five cases additional written questions were sent to the informants by letter or e-mail; however, only one person responded to these. The author also spoke to freelance insurance brokers and representatives of insurance companies. Information derived from discussions with human geneticists, medical experts in the field of HD, and the DHH chair was also included in the study. I also used notes from conversations with those living with HD conducted at the 2004 DHH annual conference. The written poll and telephone interviews were conducted between May and December 2004.

Findings

A total of forty-eight questionnaires were returned. Ten of these came from the contact group in Austria, one from the United States,[11] and the rest from all parts of Germany. A clear majority of the informants (twenty-nine persons) declared that they had already experienced (genetic) discrimination, and almost one fifth feared they would experience it in the future (nine persons). About the same number (ten persons) answered both questions in the negative. "Persons at risk" for Huntington's disease were in the minority among those who described their experiences of discrimination;

most of the respondents reported practices of disrespect, stigmatization, and exclusion based on already existing symptoms.

Like several other studies, the findings of this investigation are grounded primarily on self-reported experiences of genetic discrimination. The study shows that some individuals are convinced that they have been discriminated against due to genetic factors; however, the question of whether the reported incidences concern a practice without any legal justification, or a legitimate differentiation, cannot be decided. It is important to keep in mind that not every form of unequal treatment is morally reprehensible or legally prohibited. A company's decision to only hire applicants with the best school marks is certainly unobjectionable, whereas, in contrast, only accepting applications from men or from people with white skin color would not only spark protests but would in any case be illegal (Hellman 2003: 101–105; 2008). The question then arises of which instances of differential treatment based on genetic characteristics can be considered socially acceptable or legally justified.

However, it will be difficult to arrive at a neutral and independent assessment regarding the alleged discriminatory practices, as some scholars seem to assume (cf. Otlowski, Taylor, and Barlow-Stewart 2002: 164–167). The idea of "objective evidence," something that will make it possible to distinguish clearly and unambiguously between "fear of and actual GD, between perceptions and independently verified instances of GD" (Otlowski et al. 2012), might be mistaken because it disregards a central feature of (genetic) discrimination: what is perceived as discriminatory is itself subject to social values and normative conflicts. What one side might consider legally justified and morally legitimate practice, the other might regard as disrespectful, disadvantaging, and exclusionary. Thus, the idea that it is possible to distinguish neatly and objectively between "fair" and "unfair" discrimination runs the risk of omitting a vital part of the discussion on genetic discrimination (see Klitzman 2010).[12]

REPORTS OF ORGANIZATIONAL DISCRIMINATION AND STRATEGIES TO AVOID IT

In four cases "persons at risk" for HD reported genetic discrimination on the part of organizations. All these involved the insurance industry. A young woman, several members of whose family already live with HD, described an exemplary case.

Case 1

In 2002, the woman submitted an application for a "dread disease" insurance that covers serious diseases. The application form included questions on the state of the applicant's health, in the answers to which possible "hereditary illnesses" were to be listed. The woman disclosed that she often suffers from migraines and had poor eyesight, and also that her grandfather, father, and uncle all lived with HD or had already died from it.

The insurance company initially rejected her application without explanation. In response to her inquiry by telephone, the insurance agent told her that the insurance coverage had been denied to her because of the history of HD in her family. He regretted this decision, he said, but it was not possible for him to change it. The agent suggested that she could undergo genetic testing. If the test result were negative, there would be no obstacle to an insurance contract. Shortly afterward, the woman received a letter from the insurance company rejecting her application and stating that "unfortunately we are bound to medical statistics in our decisions and thus forced to reject your application due to the family history of Huntington's chorea."[13] This decision was—in the woman's own words—"a slap in the face" (R 39; see also R 21, R 8).

Case 2

The younger sister of the woman in Case 1 had a similar experience. She tried twice to acquire a policy from the same insurance company, and both times was rejected due to her family history (R 38).

Case 3

In the next case, the mother of the respondent was diagnosed with HD in the mid-1990s. When the daughter found out about the illness, she took out a life insurance policy that also included disability insurance as quickly as possible. She reacted—in her own judgment—in a state of "panic," did not carefully compare the offers from the various insurance companies with one another, and thus purchased a policy with unfavorable terms. After a few years, the woman decided to gain certainty about her risk of falling ill of HD and underwent genetic testing. She contacted a local HD self-help group and with their help found a psychologist who advised her during the genetic counseling and before the genetic test. The genetic test ultimately revealed that she was not a "carrier" of the mutation. With this result, the woman precipitately broke off her psychological counseling.[14]

After she had received the (negative) result of the predictive genetic test, the woman wanted to change insurance companies in order to acquire a policy with better conditions. She contacted an independent insurance broker recommended by an acquaintance, who explained to her that at present she could not take out a life insurance policy because she had recently been in psychotherapy. The independent insurance broker knew nothing about HD and the possibility of a predictive genetic diagnosis, nor was she aware that the "guidelines for the application of a molecular-genetic test"[15] strongly recommend psychological counseling during the test period and after the results of the diagnosis are known. Apparently, for the insurance broker, the reason for the psychotherapy was irrelevant. In her view, the simple fact *that* the woman had made use of psychotherapeutic counseling sufficed[16] (R 21).

Case 4

A respondent described her difficulties in acquiring an insurance policy against professional incapacity. In 2000, several insurance companies had rejected her due to her family history. Because the person wished to remain anonymous, however, it was not possible to receive a more detailed description of the case and specific proof of the rejection decision (R 8).

These cases are in line with the findings in the literature on genetic discrimination that focuses on organizational decisions. However, the picture remains incomplete. What is missing in most accounts is the diversity of strategies that individuals use to provide against experiencing discrimination. The investigation carried out by Geller and her colleagues is an exception. Even here, however, only a small portion of the study is devoted to strategies to avoid genetic discrimination: "These strategies included purchasing insurance policies before genetic testing, being tested anonymously, paying out-of-pocket for tests so that insurance companies would not obtain the results, providing partial disclosure of relevant information and, sometimes, providing incorrect information." (1996: 79)

The first study to address systematically the behavioral responses individuals develop to cope with the risk of genetic discrimination was undertaken by Bombard et al. (2007; see also 2008). The authors conducted thirty-seven semi-structured interviews with asymptomatic individuals in Canada who were found to have the mutation responsible for HD. Their findings suggested four main strategies used by individuals to manage the potential for or experiences of genetic discrimination: keeping a low profile, minimizing, pre-empting, and confronting genetic discrimination. The majority of participants chose the strategy of "keeping a low profile" to cope with the risk of genetic discrimination (62 percent). They mostly kept their family history or genetic test results private and were very hesitant to share the information concerning their risk status. Whereas some did not communicate the risk to anyone, others only spoke to a very limited group of people about it (mostly family members and close friends). The authors identified a second strategy among participants, which they termed "minimizing genetic discrimination." This is characterized by attempts not to reflect directly upon experiences of genetic discrimination or to disconnect the incidents from experiences of discrimination. Around 30 percent of the participants used this strategy. "Pre-empting genetic discrimination" is the third strategy, and this includes all attempts on the part of those at risk for HD to evade or "protect" themselves against genetic discrimination. Typical measures that are linked to this form of discrimination management include purchasing insurance policies prior to predictive testing or ensuring that one's tests results are not listed in their GP's medical files. Nearly half of the participants described using this strategy. The last strategy, employed by 27 percent of the participants, aims at "controlling genetic discrimination." It is characterized by a set of approaches which are mobilized after an experience of genetic discrimination and which seek to resist and challenge it. Individuals using this strategy confronted the perpetrator, sought advice, or rejected the basis for discrimination. Bombard and her colleagues conclude that the four strategies identified in the study provide an exploratory framework that may be useful to explain similarities and differences among individuals living with an increased risk for HD (or other late-onset diseases) in their ways of coping with (the risk of) genetic discrimination (Bombard et al. 2007, 2008).

The results of the author's own investigation confirm the importance of this perspective. Respondents resorted to modes of action and patterns of behavior anticipating "a latent rejection" (R 47), adapting their behavior to real or imagined social constraints. One participant accounted for not having experienced genetic discrimination "because I do not mention my being a 'person at risk'" (R 12). Another reported that she lived her life along the lines of, "The fewer people know, the better" (R 21). The most effective means of pre-empting genetic discrimination is not to draw the (possible) mutation carrier status to the attention of organizational actors.

This "precautionary secrecy" (R 47) can be seen in the way some individuals dealt with the question about "hereditary illnesses" asked at the official medical examinations conducted for those wishing to acquire the status of full civil servant in Germany. Some respondents provided incomplete or erroneous details in order to get the job they wanted (R 44, R 7). In the case of a female police officer, the fear of her colleagues finding out about the disease led to her reading through the medical reports on her father, who was living with HD, to discover whether they contained anything indicating that she herself was at risk from the disease. She was afraid her superiors could force her to undergo predictive genetic testing in order to ascertain whether she was fit for service: "If somebody in the force had found out, I would not have been made a full civil servant! If it were to be made known now, I would probably be forced to take the test or to give up my job." (R 19) A woman who had held a temporary position at a university mentioned similar fears. She was certain that she would have lost her job and been "bullied out" if she had told anybody in the department about her risk of developing HD (R 40).

Respondents concealed their family histories both with regard to employment and in other organizational settings. One person gave inaccurate answers to questions about her family history when she applied for a private health insurance (R 7). In another case a man who, together with his wife, wished to adopt a child kept secret the fact that members of his family lived with HD "because we were not sure whether that would have counted us out. We stated that my father had multiple sclerosis." (R 46)

This strategy even extends to choice of partner, friends, and neighborly relations. Individuals at risk of the disease refused to "come out" (R 40) and deliberately kept secret their risk of falling ill, fearing that otherwise their relationships with friends and acquaintances might change for the worse (R 44). In rare cases, even partners were not informed about the illness and symptoms involved (R 40). Many respondents conceived of not disclosing their genetic risk as a pressured constraint which affected their personal and social relations in a negative way, limiting their communicative options and forcing them to withhold from others important information about themselves and their own future. As one respondent put it, the fear of being "found out" is "a decisive factor that itself already represents a form of discrimination" (R 19).

Concealing genetic risks seems to be characteristic even of the way those threatened by genetic discrimination relate to one another. One woman pointed out that among "persons at risk" the (positive) results of genetic tests are among the "best kept family secrets." As she put it, "Among my circle of acquaintances it is actually only families who can be sure that no more of the offspring can contract Huntington's that talk openly about the disease." (R 24) The decision of those affected by the disease "not to let anything show" can also extend to their closest relatives; (grown-up) children may be excluded from discussions about the illness and their own risk status (R 47, R 40). A man living with HD broke off contact with the entire family of his brother, who was likewise affected by the disease. Apparently he did not want his children to learn anything about the disease and its hereditary nature from their relatives (R 38, R 39).

The findings show that for a systematic account of genetic discrimination it is not sufficient simply to record cases of discrimination; it is also important to consider the strategies which "persons at risk" or already symptomatic individuals use to anticipate negative categorization and the ways in which they adapt their behavior accordingly. It is illuminating to confront this "precautionary secrecy" (R 47) about genetic risk status with Erving Goffman's studies of stigmatized persons. Goffman differentiated between two forms of stigma:

> [D]oes the stigmatized individual assume his differentness is known about already or is evident on the spot, or does he assume it is neither known about by those present nor immediately perceivable by them? In the first case one deals with the plight of the *discredited*, in the second with that of the *discreditable*. (Goffman 1986: 4; emphasis in original)

Individuals with a genetic risk of falling ill are "discreditable," and the crucial problem for them is to filter and manage information about their "genetic differentness" so that they do not end up as a discredited person. The resulting permanent "information control to manage genetic peculiarities" (Scholz 1995: 52) typically consists of a double strategy. On the one hand, "persons at risk" avoid situations in which knowledge of their own "genetic errors" could become relevant by practicing careful secrecy (especially towards organizational actors). On the other hand, "genetically discreditable" individuals share their risk status with a smaller group of people (partners, family members and close friends). In this way, not only is such a status loaded with social meaning, but new constraints and feelings of guilt are also triggered: "Close relationships to others, who through mutual revelations of invisible defects are repeatedly confirmed, force the individual either to admit the situation to the trusted person or to feel guilty because the individual does not do this." (Ibid.: 53)[17]

"HITLER'S LONG SHADOW": MECHANISMS OF INDIRECT DISCRIMINATION

In addition to discrimination against individuals who are directly confronted with disadvantages or disrespect, the study also revealed forms of "indirect" discrimination. Reproductive choice is a key arena in which indirect mechanisms of discrimination are experienced. Here the central question is whether or not persons who are already living with HD, or may do so in the future, should have children as they might also be affected by the disease. As the reported experiences show, in Germany this question cannot be separated from the country's history of eugenics. During the Nazi era in Germany, the "Law for the Prevention of Hereditary Diseased Offspring" led to the systematic registration of entire families with HD, obligatory reporting by physicians, forced sterilization, and later the murder in extermination facilities of those affected by the disease.[18] "Hitler's long shadow" (R 7) is still cast over the present, because there are many patients with HD and "persons at risk" who lost their relatives as a result of the Nazi extermination policy. Some respondents actually fear that the eugenic ideas of the past will—albeit in a different form—resurface in the future. For one man who is already symptomatic for the disease, the way society deals with disabled people today is reminiscent of the practice of euthanasia under Nazi rule: "If the fact that disabilities can be filtered out is generally rooted in our minds, nobody will understand why there are still disabled people. In the past, Hitler already planned to eradicate disabilities by means of euthanasia and castration." (R 3) That such worries are apparently not completely unfounded is demonstrated by the following case in which one physician emphasized the "model character" of National Socialist eugenics.

Case 5

A woman whose husband lives with HD spoke to the hospital physician treating him about the implications of the disease. She reported that the physician explained the significance of the disease in the following words: "This is a hereditary illness which unfortunately is incurable and has to be exterminated.

Horrible tragedies took place in the past. One should not be allowed to have children, but only adopt them. The Nazis sterilized them all, and it was the best thing they did. Because the families were all stigmatized." The physician advised the woman to annul her marriage and put her husband in a nursing home. For the wife this was "a terrible experience," "a trauma," and a "serious injury." Shortly thereafter, the medical authorities turned down the man's application for admission to a health resort for rehabilitation purposes. The woman appealed against the decision and was informed by the person who had delivered the medical expertise: "What is someone like that going to do at a spa? This is far too expensive, and does not make sense." (R 22)

In contrast to this experience, some respondents expressed their conviction that the current treatment of ill and disabled people has nothing to do with the eugenic practices of the Nazi regime (e.g. R 16). They stressed the historical difference to the way HD patients and "persons at risk" were treated during National Socialism. However, there is a striking similarity between the two positions. Whether they see a caesura or continuity, a common assumption is shared by supporters and opponents of the argument that eugenic practices of the past determine the present. In both accounts eugenics represents a politics that functions primarily via repression, constraint and violence.[19]

A different account is to be found in other answers. Here eugenics is interpreted less as a direct constraint that might even lead to physical extermination, and more as indirect guidance of individuals shaping their preferences and enlarging their options. Instead of eugenic programs enforced by state institutions relying primarily on repressive means, this concept of eugenics captures measures and practices to enhance individual well-being, self-determination, and choice. Whereas classical eugenic programs aimed at improving the quality of the human race or the national stock by steering hereditary processes, "liberal eugenics" does not address the population at large but aims to maximize individuals' health and reduce their suffering. It proposes to "arrive at the best internal and social arrangements by allowing informed prospective parents to be guided by their values in choosing enhancements" (Agar 1998: 139; 2004; Gavaghan 2007).

In this context, "persons at risk" for HD see themselves confronted with changed social norms and institutional expectations that do not exclusively focus on explicit restrictions and prohibitions (see, for example, Shaw 1987), but foster and encourage a "mature," "responsible," and "risk competent" attitude to health issues and reproductive decision making.[20] Some respondents feared increasing pressure on possible "carriers" to undergo predictive genetic testing in order to avoid health risks to oneself and one's offspring. One respondent, for example, was worried "that 'persons at risk' could be forced to undergo diagnostics and that the possibility of prenatal testing reduces the level of acceptance of disabled people and restricts the development of effective medication and therapies" (R 17; similarly R 21). Several informants stated that due to their risk of falling ill and the hereditary nature of the disease, they had consciously decided against having children or "ignored the issue" (R 47). In their view, this "voluntary renunciation" (R 7) is necessary, because it would not be responsible in their opinion to have a child that could possibly later be affected by HD (R 23).[21]

In two cases, "persons at risk" reported that doctors had indicated to them that the decision to have biological children was irresponsible and morally questionable. In one instance, after a male patient had a positive result in his HD test, the doctor stated categorically to him and his wife: "It is up to you to ensure that you have no more children and that the children you have already will have no children themselves." The couple was then provided with the telephone number of the nearest genetic counseling service (R 45). In another case, a woman at risk of developing HD described how a consultation when she was pregnant "had quite clearly tended towards an abortion" (R 44). Even though both these instances occurred more than twenty years ago, it is not clear that the principle of non-directiveness in genetic counseling is currently always treated with due respect. This is highlighted by the findings of a German study on the psychological and social aspects of predicting Familial Adenomatous Polyposis—an inherited disorder characterized by cancer of the large intestine (colon) and rectum. A series of surveyed patients reported that their counseling physician attempted to exert influence on their family planning. For example, one of the surgeons said: "Please refrain from having children; otherwise you will only transmit the disorder." (Schmedders 2005: 8; 2004)[22]

"NOT ACCEPTABLE": CONTRADICTORY EXPERIENCES OF INTERACTIONAL DISCRIMINATION

Many "persons at risk" and those already living with HD complained about the general lack of knowledge about the genetic etiology of the disease. According to them this was due, among other things, to the fact that the disorder occurs relatively rarely and is little known. Observers often ascribe the resulting motor ailments and speaking difficulties to drug abuse or alcohol problems. Those affected deplore these prejudices and the lack of understanding with which they are often confronted; people often "look at them disapprovingly," see them as alcoholics (R 27, R 29), mock them (R 33, R 34) or even insult them (R 30). Interestingly, respondents reporting these incidents claim that they are genetically discriminated against. This evaluation does not fit the classic definition of genetic discrimination because, in the cases just mentioned, the discriminatory practices apparently refer to the manifest symptoms of HD. The observers obviously mistook those symptomatic with HD for drug addicts or alcoholics, probably assuming that they were responsible for their behavior. However, respondents did not make a strict distinction between discrimination against disabled individuals and the chronically ill, on the one hand, and discrimination on the basis of the genotype, on the other, but rather stressed the continuity of discriminatory practices. As one respondent put it, criticizing the focus of the study on genetic discrimination for being too narrow: "Why *genetic* discrimination? What is at stake is discrimination *in general*." (R 5; emphasis in original)

It comes as no surprise that many of those already symptomatic for HD experience the same problems chronically ill and disabled people often face in their daily lives. Respondents deplored the fact that many people had "problems and difficulties" dealing with them (R 39)—that, having found out about the disease, they subsequently avoided them (R 23), expressed a lack of understanding for their suffering (R 27), or responded "almost maliciously" when they encountered a person living with HD (R 40, R 47). Respondents registered increasing social isolation because friends and acquaintances often withdrew, broke off contact, and avoided individuals in whose families cases of HD had occurred (R 27, R 32, R 38, R 39, R 42, R 47).

The experiences of disrespect and stigmatization, as suggested above, extended to the families of those affected. Instead of reacting with sympathy and practical solidarity, contact with symptomatic family members was often reduced or even completely broken off. In other cases, responsibility for HD was ascribed to the person affected by the disease and the physical and psychological problems arising were denied (R 40). In one case, for example, the sister-in-law of a man living with HD accused him of "pretending" when he had to give up his profession due to the disorder. She alleged that she had to work so that he and her sister "could lead a comfortable life." Disapproval also came from the husband's family. Relatives made a taboo of the disease and its hereditary nature: "In his family no one spoke about Huntington's disease. They regarded us as traitors since we said what [name] had." (R 22)

However, whereas some participants experienced disrespect and exclusion because they were held responsible for their disease or because they were not regarded as ill at all (but rather as drug addicts or alcoholics), others report that they were treated differently compared to other patients or people with disabilities. They felt confronted with an implicit hierarchy of diseases, HD representing an especially dangerous kind. This issue is addressed in the case of a young woman who developed HD at a very early age and who held a voluntary position on a children's ward until her superiors terminated her contract.

Case 6

At the beginning of 2003, the woman applied to the children's charity of a German university hospital for a voluntary position as a social worker for sick children. The disease was still in its early stages, such that the woman was in no way mentally impaired, as a leading German HD expert confirmed in 2004. On taking up her position, she informed her superior of the HD diagnosis. Because her superior was a paraplegic and used a wheelchair (she also recalled that he suffered from occasional cramps), she had no fear of speaking openly with him about her own disease.

The relationship with her superior, however, quickly deteriorated. He indicated to her that she was "overtaxed" and after some time forbade her to be alone with the children in a room (other volunteers were allowed this). This measure resulted, among other drawbacks, in her not being able to work when her colleagues were on holiday. Because she found this to be an unfair restriction, she asked for a meeting with her superior. During their discussion, he explained that he could not assume "responsibility" in case something happened to the children. He claimed that because of her disease it could not be ruled out that she could drop a child or would not be able to run after a child.

Annoyed that she was being treated differently from other voluntary helpers, the woman appealed to the head of the institution. The director explained to her that the superior had acted correctly, because she was supposedly "psychologically unstable." The director told the respondent that she was a "danger" for the children because it was unclear how she would behave in the future. As a person affected with HD, she was "unpredictable" and thus "not acceptable" in the hospital. The young woman offered to present medical reports to corroborate that she was physically and mentally completely capable of taking care of children. The director, however, rejected the proposal and terminated the contract (R 23).

The justification offered for the woman's lack of capability or even the "danger" she was said to pose is all the more surprising because her superior, who without a doubt was physically restricted, took care of the children alone without having a second person present. Moreover, the voluntary colleagues of the young woman were in the majority retirees, many over seventy years old, who openly admitted to the respondent that they often felt overtaxed by the children. Because at the time she worked in the hospital the woman did not suffer from any marked physical or mental limitations, questions arise concerning the grounds for her rejection by the hospital's management. On the one hand, it may be that ignorance about the disease itself, its highly complex causes and exceptional symptoms, was responsible for the woman being treated differently. On the other hand, her superiors' fear could well have been rooted in the specific genetic nature of the disease, which in their own words they regarded as "unpredictable" and "dangerous," to the extent that they forbade the young woman to be alone with the children.[23]

In the following case, the care workers in a secure psychiatric hospital had similar fears of a man symptomatic for HD who had been admitted following a series of suicide attempts.

Case 7

According to his daughter, despite dysfunctions in physical coordination the patient in question was not aggressive and mentally "very alert." As the woman recalled, the hospital staff had had no experience with HD and displayed no interest in finding out about it. Because of the uncoordinated manner in which the patient moved, the care workers refused to cut his fingernails. These grew to such a length that the man injured his face. He was also insulted by several care workers ("Here comes our dancer, wait until he falls on his face and starts to cry again"). When the daughter pointed out to the responsible physician the improper behavior of the personnel, she was told that the care workers were "afraid" of her father, supposedly because his "disease is so strange." According to the doctor, they feared the patient would lash out at them ("getting a clip round the ears") (R 19).

These reports seem to indicate that some of those symptomatic of HD encounter specific fears that could possibly be linked to the genetic "nature" of the disease. However, it should be noted that HD is a degenerative neurological disease with peculiar symptoms: it is different from many genetic illnesses that manifest themselves "only" in physical restrictions and limited capabilities. It is possible that the experiences with disrespect and disadvantage described here are the result of a fear of actual or imaginary behavioral and personality changes of the patients. In this context, further studies might focus on whether those with non-genetic degenerative neurological illnesses are subjected to similar treatment or whether there are significant differences between the two groups of patients.

Those who suffer from genetic discrimination perceive it as particularly disrespectful and disparaging, because genetic factors are considered to be especially powerful in everyday discourse. Genetic information seems to be suited to revealing hidden truths about individual health risks, physical or mental characteristics, and personal identity. The social exceptionality of genetic knowledge is confirmed by the results of a qualitative study conducted by David Armstrong, Susan Michie, and Theresa Marteau. These authors analyzed sessions of genetic counseling and investigated how the identity of the counselees is constructed in the counseling process. The empirical analysis shows that those affected by genetic diseases differ from other sick persons in one fundamental area:

> Being diagnosed as having a disease or having a spoiling surgical procedure . . . means a *new* identity, one that is added to or conflicts with an existing identity. In genetic disease there may still be stigma, both felt in terms of feeling 'spoiled' and enacted (for example, from insurance companies), but it has a different basis. Genetic disease differs in as much as it promises to reveal who the individual always has been, not a new addition but a revelation about an underlying identity that had been concealed. (1998: 1657–1658; emphasis in original)

The medicalizing gaze informed by genetic knowledge shapes even family and kinship relations. In an important study published some years ago, medical anthropologist Kaja Finkler analyzed the impact of the new genetics on family relations and kinship. Her work is based on extensive fieldwork in the United States and interviews with breast cancer patients, healthy women from families affected by breast cancer, and adopted children seeking contact with their biological parents. The results show that the ideas of her interviewees on family and kinship are dominated by a "genetic inheritance ideology" (Finkler 2000; see also Finkler, Skrzynia, and Evans 2003). Concepts of inheritance and the significance of genetic transmission informed the experiences of the individuals and their relations to their children, parents, and other relatives. Finkler demonstrates that family and kin connections are framed in terms of genetic ties and in view of possible genetic risks. In this perspective, families are less defined by voluntary bonds than by a common genetic heritage. The genetic map postulates proximity between individuals separated by spatial or genealogical distance. It transcends emotional alienation and conflicts between family members. Even family members who no longer see each other or who have broken off contact remain genetically close. In this perspective, family ties are primarily experienced in terms of genetic bonds that are seen as more fundamental and stable than family ties based on love, affection, and the free will of the partners.

From a cultural point of view, genes are often seen to symbolize something fateful and unalterable (Nelkin and Lindee 1995; van Dijck 1998). Genetic risks are represented as unique in that they neither enter the body from without nor do they destroy the body's defense mechanisms from within. Rather, they are part and parcel of the individual and not—as with other health risks—a temporary risk that can be treated and eradicated.[24] The deterministic picture of genetic factors often prevailing in the media and everyday discourse spurs feelings of helplessness and fear. He or she not only "carries" or "possesses" genetic risks, these are seen as an integral part of a person's own physical existence (see Kavanagh and Broom 1998).

Before turning to the normative implications of genetic knowledge in the next chapter, I will first examine the limits of this study and point out difficulties in the interpretation of the data.

EXTENDING THE CONCEPT OF GENETIC DISCRIMINATION

The findings of this study reveal the limits of a narrow concept of genetic discrimination. Focusing on organizational actors such as insurance companies, employers, and adoption agencies, previous studies of genetic discrimination tended to marginalize important fields and dimensions of genetic discrimination. First, they did not take into account the plurality and diversity of strategies that "persons at risk" use to pre-empt genetic discrimination. As the findings of this study suggest, it is not sufficient just to focus on "cases" of genetic discrimination. Studies should also investigate what strategies individuals employ to counter what they conceive of as a threat to their private and professional life.

Second, because studies of genetic discrimination have so far concentrated on mechanisms of organizational discrimination, they were only sensitive to cases where particular individuals were directly confronted with disadvantages and disrespect. Verdicts of social unworthiness, prejudice structures, and stereotypes that often provide the normative basis for organizational decision making and discriminatory practices have been left outside the analytic frame. This "indirect" discrimination includes all those factors that limit the choices and options available to "persons at risk."

Third, the findings of this study suggest that it might be useful to distinguish between two levels of genetic discrimination: one that is based on personal interaction, and another that builds on organizational settings. The former includes more or less spontaneous stigmatization or disrespectful words or acts (individual or collective) against other persons, whereas the latter focuses on durable organizational structures. Although these two levels can only be separated analytically and in reality often overlap, the distinction makes it possible to investigate the conflicting and contradictory experiences of discrimination suffered by those affected by HD. Whereas some are taken to be alcoholics or drug addicts and as such not regarded as ill at all, others are addressed as patients deserving some kind of exceptional treatment because they represent a particular danger to themselves and others.

Two implications for further research result from these findings. On the one hand, it is necessary to determine whether genetic discrimination can be empirically shown to represent a form of discrimination *sui generis* that needs to be analyzed separately from discrimination experienced by disabled or chronically ill persons. On the other hand, there is a need to develop a theoretically informed concept of genetic discrimination and to investigate how it differs from related notions like "geneticization" (Lippman 1991) or "geneticism" (Wolf 1995).

However, given the low response rate and considering the fact that there are approximately 8,000 people in Germany living with HD and many more "persons at risk," it is obvious that this study cannot be a representative survey. In addition, one must take into account that, given the high penetrance of the disorder and the exceptional predictive value of the genetic test, HD occupies a special position within the spectrum of genetic diseases. Individuals with other genetic properties may be confronted with forms of genetic discrimination that are significantly different from those analyzed here. The results of this exploratory study only allow cautious conclusions to be drawn about experiences of genetic discrimination. Future studies will have to determine more precisely, and in greater detail, the prevalence and the specific forms of genetic discrimination in Germany.

5 The Regime of Truth and Dimensions of Genetic Responsibility

> Our genes are the most intimate biological possession we have. How we deal with them is our personal responsibility. (Schmidtke 1997: 13)

> The construction of health as a moral issue has generally been confined to discussions of voluntary health risks—people's lifestyle choices or behaviour. More recently it has been observed that individuals not only have a responsibility to avoid voluntarily exposing themselves and others to health risks, but also may be seen as bearing some responsibility for their genetic risks . . . Indeed, it can be argued that because genetic risks are portrayed as part of the individual's make up their responsibility to act to protect their health, or the health of future generations, is emphasized, for inherited risk can not be blamed upon external sources. (Hallowell 1999: 98)

In the early 1990s and at the beginning of the Human Genome Project, the Canadian sociologist Abby Lippman coined the term "geneticization" to account for the social and cultural impact of the new genetics (Lippman 1991). Lippman employed this neologism to critically analyze a (medical) perspective that conceives of genes as a program for the development and regulation of organisms and regards genetics as the central conceptual model to explain human life and behavior, health and disease, normality and deviance. Since its original formulation, the term has been used by many social scientists to designate "a social process through which concepts, theories, social structures, and individual and social practices . . . are gradually changing to recognize and incorporate the explanations and rationalities of the new genetics" (Ibid.: 19; see also Fitzgerald 1998; Koch 2002; ten Have 2004). Although some theoretical and normative underpinnings of the concept have met with significant criticism (Gibbon 2002; Novas and Rose 2000; Rouvroy 2008), it is still useful as an analytical instrument and descriptive concept (Kollek and Lemke 2008)—even if the "molecular vision of life" (Kay 1993) has lost a lot of scientific credit since the heyday of the Human Genome Project and has been replaced by a more complex understanding of genes and genetics.

The concept of geneticization is part of a series of objections that have been raised to counter genetic reductionism, focusing above all on the "unscientific" and "ideological" character of such arguments. Scholars from the humanities and social scientists have rejected the "gene myth" (Hubbard and

Wald 1997) and pointed to the significance of social mechanisms, psychological factors, and historical knowledge (Nelkin and Lindee 1995; Duden 2002). Most of the criticism has focused on geneticization as a process which rests on a "false," "distorted," or "one-sided" form of scientific knowledge. Indeed, this line of critique remains pertinent even within the more recent appraisal of complexity in genomic research and the emergence of new disciplines like systems biology and epigenetics. As we saw in Chapter 1, the new approach enacts a rather selective and limited account of gene-environment interactions which still privileges localizable genes over environmental and social factors. The latter are "reduced to a set of covariates or confounders that need to be statistically controlled for rather than becoming an object of investigation in themselves" (Bauer 2009: 128). Therefore, the new narrative has justifiably been called a form of "neoreductionism" (Lock 2005: S48) or an "enlightened geneticization" (Hedgecoe 2001).

However, it does not suffice to rebuke reductionist explanations and under complex concepts in genomics for being one sided, ill informed, or geared only to specific interests. In addition to this indispensable form of critique, it is necessary to analyze the practical implications of this reductionism and the social consequences it will have beyond all scientific debate. In this respect, it is important not only to focus on the "falseness" of genetic reductionism, but also to analyze how it produces truth. Theoretical attention must be focused on "spaces of representation" or "regimes of signification" (Rheinberger 1997b; resp. Kay 2000: 18) which organize an epistemo-political field of the visible and the expressible, controlling the diverse forms of signification and defining the conditions for truth and falsehood. What is at stake is a representation of biological processes which guides practical interventions and structures the terrain for possible action. In other words, instead of juxtaposing reality and reductionism, we should study genetic reductionism as a "truth regime."[1]

In this chapter, I will argue that this truth regime is closely connected to mechanisms of indirect discrimination. My thesis is that the idea of genetic medicine cannot be separated from the techniques of governing others and the self. These governmental techniques are employed to contribute to the "rationalization" and "responsibilization" of health behavior by producing information for individuals and families so that they can predict genetic risks and—if possible—prevent diseases. In the future, genetic conditions could increasingly be handled as risk factors, establishing new areas of (individual) decision making as well as hitherto unknown moral responsibilities.

GENETIC MEDICINE AS NORMATIVE ETIOLOGY

The analysis of genetic medicine as a "truth regime" proposes a nominalist approach combining two lines of investigation—one archeological and the other genealogical.[2] The archeological analysis demonstrates that genetic explanations or models of disease possess a certain basic form which connects a set of signs and symptoms with an underlying genetic cause. This etiological model has only been established in recent years, and it neglects or marginalizes alternative forms of disease causation. The question arises of which conditions or characteristics are grasped today in genetic terms that were not always regarded as genetically determined or caused, and what effects this form of knowledge has (e.g. on medical practices, health policy, scientific research, commercial interests). This does not necessarily mean to negate or question the validity claims of genetic knowledge. The point at issue here is less the (juridical) question of whether certain disorders are correctly defined as genetic diseases than the investigation of the historical conditions of emergence of a specific medical knowledge.

Edward Yoxen explains this perspective in the following words:

> I take the view that the origination, development, and utilization of particular forms of knowledge, for example, about those states of conditions we call diseases, are socially determined and have to be accounted for in sociological terms . . . One can claim, as I do, that many of the phenomena of genetic disease are grounded in a material reality while, at the same time, asking why we isolate or delineate certain phenomena for analysis, why we say that they constitute diseases, and why we seek to explain their nature and cause in genetic terms . . . In seeking to consider the sociology of genetic explanations of disease, we are essentially asking what kinds of evidence and explanation persuade. (1984: 41–42; see also Lemke 2003)

The second—genealogical—step in the nominalist method builds upon the first. When certain concepts of disease are becoming problematic and start losing their "self-evident" or "natural" status, it is indispensable to uncover the network of "connections, encounters, supports, blockages, plays of forces, strategies and so on, that at a given moment establish what subsequently counts as being self-evident, universal and necessary" (Foucault 2000: 226–227). The ensuing questions are, who "profits" and how from the "geneticization of medicine" (in terms of, for example, financial gain, political influence, scientific reputation, social prestige), and which structural constraints and institutional practices are coupled to this prevailing understanding of disease?

This form of analysis takes into account the performative and strategic character of genetic medicine, which is either ignored or discussed too briefly in the scholarly debate on genetic technologies and their medical implications. At present, genetics plays a role in diagnostic applications, a much more limited one in disease prevention, and hardly any in therapy. This rather limited impact of genetics on clinical practice is further restricted to industrialized countries with advanced medical services and technological infrastructures. Nevertheless, the concept of genetic disease contributed decisively to mustering the financial and intellectual resources for the Human Genome Project. The mobilization of the concept paved the way for technical innovations that did not exist at the beginning of the project.[3] At the same time, media coverage and the popularization of the idea of genetic medicine mobilized support and rendered it socially acceptable.

> [T]he concept of genetic disease, enthusiastically appropriated by the medical sciences for complex institutional and economic reasons, represents an ideological expansion of molecular biology far beyond its technical successes. I also want to argue that the general acceptance of this concept has, in turn, proved critical for the direction that subsequent technical developments in molecular biology have now begun to take. (Keller 1992: 293)

In addition to this performative dimension, the strategic orientation of the "genetic point of view" (Childs 1977: 8) has to be examined, taking into consideration the structural constraints and institutional mechanisms within the health system.

> Health-related conditions have to be packaged so as to fit the institutional, professional, and conceptual structure of modern medicine. In the case of conditions believed (rightly or wrongly) to be genetic in cause, the mechanistic basic explanatory form of a 'genetic disease' has been constructed to fit the contemporary context. This is not to say that our understanding of particular conditions is necessarily invalid, misleading, or the result of deliberate misrepresentation, but it is to say that medical conceptualization of particular conditions bears the mark of a struggle over whether and how to think about them. (Yoxen 1984: 48)

The rise of genetics within medicine and the emergence of the concept of genetic medicine cannot only be ascribed to internal structures of the medical complex or professional policies by geneticists, but must be analyzed in the context of broader social changes.[4] The concentration on genetic dimensions of disease causation (and the simultaneous neglect of other factors) is reducible neither to a dynamics inside science nor to a result of medical and technical progress. Hagen Kühn's idea of a "normative etiology" is useful in this context, as it stresses that the career of certain paradigms of disease depends not least on whether they "fit" into more general conceptual and political strategies. Whether particular concepts of health receive political consideration, gain social acceptance, or appear scientifically respectable depends just as much on social mechanisms of selection as on political interests: "Here 'survival of the fittest' does not refer to the selection of the strongest or the most appropriate, but of the best adapted to the status quo." (Kühn 2000: 13; see also Tesh 1988)[5]

Kühn's argument is that the "conditions of survival" and strength of self-assertion of etiological concepts depend decisively on power relations in society. Without questioning existing scientific explanations of disease, he claims that some concepts of disease and models of intervention have a better "chance" within a given society of being considered and incorporated into health policies. These chances of being selected do not derive from their internal consistency or scientific truth, he maintains, but are rather due to their "correspondence" with external imperatives and constraints resulting from the existing social order. From this perspective, concepts of disease not only have to fight against scientific errors and epistemic obstacles, but also and above all against persisting forms of domination and power structures.

Proceeding from this assumption, Kühn discusses different concepts of disease within the framework of an evolutionary model according to their inherent "survival probability." He ascribes to concepts that concentrate on isolatable germs and genetic conditions the best chances of being recognized and accepted, because in this case the causes of disease are situated in the area of the micro-events of the individual body—at the same time leaving the broader social context untouched. For this reason the geneticization of disease

> can to a large extent be traced back to the promise that here disease could be fought without disturbing the social and political status quo. What is more, the status quo is strengthened by the market-conform satisfaction of health needs, by new markets for investors, and by the withdrawal of legitimation for interventions in the social context of health and disease . . . On the second level of the etiological hierarchy are those causes of disease that are directly regarded as *reflective, i.e. related to oneself, 'wrong behavior' of the individual*, in other words the subjective risks of the so-called lifestyle (lack of physical movement, incorrect and excessive eating, too much alcohol and tobacco, the wrong social relationships). (Kühn 2000: 14–15; emphasis in original; cf. Kühn 1993: 131–133)

Beyond the two privileged levels of the etiological hierarchy, Kühn distinguishes additional exploratory models which are less likely to be taken up in health policies. These focus on factors of the psychological and social environment and their contribution to the causation of diseases and their development. Intervening here, he claims, is especially difficult: "Nowhere else is money more scarce, the costs of legitimation higher, and the burden of proof more difficult than in attempts to intervene in the health aspects of areas such as labor, transport, or living conditions or in social problems such as poverty, unemployment, or powerlessness." (ibid.: 16)[6]

From this perspective, a genetic understanding of disease fulfils two key functions, which decisively contributed to the emergence of the idea of genetic medicine. First, genetic medicine allows pathogenic factors to be localized in the individual, without systematically addressing the broader social and physical context of the pathogenesis. Second, the idea that genetic factors play a significant role in the development of many common diseases makes it possible to pursue the vision of a "personalized" medicine, that is to say, one focused on the individual genome. Genetic medicine not only promises to overcome diseases without changing pathogenic factors beyond the genetic level; it also proposes market-conform "problem solutions" in the form of new pharmaceuticals, diagnostic tools and preventive measures.[7]

Kühn's general argument is very convincing in stressing the intimate links between etiological discourses and power relations. However, his analysis is limited in two respects. First, it is too static, because the focus is on the "adjustment" of concepts of disease to an already existing status quo. The consequence is that the performative dynamics, in other words, the question of to what extent discourses and practices themselves constitute and change the status quo (instead of merely existing outside it and reproducing it), is not taken into account. Second, the scientific dominance and the social acceptance of the current focus on the genetic causes of disease are based less on a hierarchical arrangement of levels than on a strategic coupling of the first and second level, which functions via the concept of (genetic) risk (Lemke 2000, 2004b, 2005). Potential causes of disease are not only localized in the individual's genetic make-up; beyond this, individuals are also expected to control genetic risks just like other health risks by means of corresponding changes in behavior. Seen from this perspective, genetic knowledge provides information which should serve as the foundation for "rational" health behavior. The contours of "genetic responsibility" become more and more visible, taking up and promoting political calls for a stronger role for individual responsibility and self-care in health policies (Lupton 1993; Bause 2000; Petersen and Bunton 2002; Bunton and Petersen 2005; Rose 2007).[8]

THE DISCOURSE OF GENETIC RESPONSIBILITY

The contemporary discourse of genetic responsibility has two conditions of emergence, one scientific-technological and the other socio-political. Both stem from the 1970s and cannot be separated from one another historically or systematically. The first refers to the growing genetic knowledge and the technological options based on it that became apparent at the beginning of the 1970s. This was an important decade during which molecular-biological techniques were established within clinical medicine. In the United States and other industrialized states, genetic knowledge was used in screening programs for certain diseases, new reproductive technologies were developed, and prenatal diagnosis became part of medical care for pregnant women (Duster 1990; Rapp 2000; Wailoo 2001).

The crisis of the Keynesian welfare state in many countries and the successful implementation of neo-liberal programs from the mid-1970s onwards mark the second line of development. The massive financial support and social acceptance of human genetic research in the 1980s and 1990s were part of a broad transformation which increasingly individualized and privatized the responsibility for social risks (Robertson 2001; Galvin 2002). The establishment of "apparatuses of insecurity" (Lemke 2011: 41–55) complemented and gradually came to replace traditional mechanisms of social security. Although individuals are now addressed as prudent and cautious subjects that should choose a responsible and rational, which means risk-minimizing lifestyle, they are simultaneously incited to entrepreneurial action as risk-taking is transformed into a public virtue.[9] This applies not only to the area of health but also to provision for old age, crime prevention, occupational incapacity, and unemployment (Castel 1991; O'Malley 1996; Henman 2004; Miller and Rose 2008). The "withdrawal of the state" is accompanied by calls for self-responsibility and self-care, and the establishment of self-regulatory competences in individual and collective subjects (Rose 1999; Cruikshank 1999). Indeed, it can be stated that "the active management of responsible choice in risk taking is at the core of neo-liberal governance" (Ericson, Barry, and Doyle 2000: 553).[10]

The notion of responsibility for one's own health takes shape within this general political constellation. Since the mid-1970s, scientific and medical interest has shifted from environmental dangers such as industrial poisons, pollutants, and social factors such as stress, employment, and housing conditions to individual lifestyles supposedly responsible for most widespread diseases (Knowles 1977; for critical views: Crawford 1977; Kühn 1993; Niehoff 2002; Crawshaw 2012). Tobacco, alcohol, lack of movement, and improper nutrition became the focus of medical interest; they were later joined by genetic factors. The contemporary discourse of genetic responsibility is now part and parcel of this comprehensive political-epistemological transformation.

A Search on the Internet

An Internet search using the words "genetic responsibility" demonstrates how the discourse has gained currency in the last ten years. In 2003, a search with the search engine Google *(google.com)* produced about 120 hits for "genetic responsibility" in English, 20 for the German equivalent "genetische Verantwortung" (*google.de*). Nearly ten years later, the same Internet search showed 6,400 hits (685 German hits). As many entries were listed several times, the number of references is actually lower.[11]

Some examples from Internet research provide a good overview of the manifold uses of the term, indicating the spectrum of the discourse of "genetic responsibility."[12] Among the listed links is a report on a conference on "Genetics at the Workplace," which took place in the state of Colorado (United States) in 2000.[13] In his talk, Richard Sharp, at that time a bio-ethicist at the National Institute for Environmental Health Sciences, discussed the problem of "genetic responsibility," which he argued resulted from the increase in genetic knowledge: "Once we know that an individual is genetically susceptible to certain disorders or diseases, is there a responsibility for that person to try to stay as healthy as possible? Are individuals obligated to not take a job that might put them at risk of getting sick? What obligation do employers have to protect their genetically vulnerable workers?"

A similar line of reasoning is taken by Leonard Fleck in an early article from 1997 with the title "Genetics and Ethics: A Community Dialogue."[14] Fleck states, "We are beyond the age of genetic innocence. We are entering the age of genetic responsibility . . . Our task is to make those choices in a reflective and responsible manner that enhances and preserves our common moral and political heritage." Fleck describes in his essay the establishment of "community forums" (a type of town meeting) in the state of Michigan (United States), whose task was to discuss ethical problems posed by the use of new genetic and reproductive technologies by evaluating various scenarios. The results of the discussions summarized by Fleck include the following recommendation: "Any individual's right to privacy should be limited when it adversely affects the equally important rights or interests of others. There are some genetic facts that we are morally obligated to reveal to potential spouses and sometimes siblings, but generally not to employers or insurers."

In the third example, the term already appears in the title: "Genetic Responsibility: A Father's Part" is a short film produced in 1995 and distributed by the Canadian Broadcasting Corporation.[15] The summary and short description of the film reads, "Can a father's health and behavior affect the health of his children? Research is indicating that drug use, chemical therapies, cigarette smoking etc. can lead to problems during pregnancy, birth defects and childhood cancer." This framing of the problem is interesting because it is apparently assumed that mothers are generally "genetically responsible" for their descendants; the focus is on improper behavior by future fathers that might lead to health problems for their children later in their life. This raises the question of potential gender asymmetries in the ascribing of genetic responsibility. Are women faced with higher demands as regards genetic responsibility?[16]

Although the examples cited so far have concentrated on the moral aspect of genetic responsibility, a further semantic dimension can be identified in other texts. The focus here is less on guilt and morally questionable behavior or the confrontation of rights and duties, and more on causality and competence. One example is provided by an article in the magazine *Maxim*. The March 2001 issue contains a text under the heading of "World o'sex" dealing with the "sensitivity card."[17] The author Lisa Lombardi cites several case stories to illustrate the claim that men who are "too sensitive" encounter adverse reactions from women. The twenty-five-year-old Cindy describes her ambivalent feelings when she was seeing a man who could cook unusually well: "I dated this guy who thought cooking was art. He'd say things like 'I haven't flambéed the cherries jubilee or massaged the chicken yet.' It was too weird dating someone who's better in the kitchen than I am." Why should this be a problem for women, one that burdens their relationships with men? The explanation of the assiduous advice giver: "Women feel vestigial genetic responsibility to be good cooks. Don't mess with it. (Think how you'd feel if she rewired your stereo)." The author states that cooking is a woman's domain, and if a man wants to excel there he will find little favor. The talk of genetic responsibility in this case refers to a question of competence—a competence allegedly fixed once and for all in the genetic code and that today still determines the division of labor between men and women.

A further case example is a page from *Science Generation*, a multi-lingual Internet portal set up by the European Commission presenting bio-scientific issues for a broader public.[18] *Science Generation* is dedicated to scientific education and informs readers that genetic determinism does not do justice to biological and social reality. The author stresses that the idea that genes are the "culprits" or the "wrongdoers" responsible for a variety of negative characteristics and diseases disregards the role of environmental factors:

> Responsibility is shared 50–50: it is true that we are 'made' by our genes, to a certain extent, but they are not the only factor in play. Without our environment, education, social contacts, we would probably not be what we are . . . Genes responsible for susceptibility to certain diseases are no more than signs of a relative risk. Attempts to reduce disorders as complex as mental disease, or behaviour such as alcoholism to genetic predisposition could lead to serious abuses and to people assimilating patients with their diseases, to the exclusion of other alternatives. Fortunately, the idea of the 'bad gene' is rapidly losing support, whether the hardcore advocates of exclusive genetic responsibility like it or not!

The text is interesting because it briefly presents the moral semantics (genes as "culprits," the "accused," the "wrongdoers"), in order to defend a type of "enlightened genetic discourse" (see Hedgecoe 2001). Responsibility for certain characteristics or diseases lies neither in environmental factors alone nor is it exclusively caused by genetic factors; both contribute in equal parts to their emergences. From this perspective, genes do not possess a fateful power; quite the opposite, environmental factors make it possible to control and limit the consequences of "defective" genes.

In summary, it can be said that the term "genetic responsibility" in everyday talk and in the discourse of experts is used in two ways that are analytically distinguishable from one another: a scientific complex of causality/competence and a moral complex of guilt/duty. To speak of "genetic responsibility" can mean, on the one hand, that genes cause X or Y, that they are the "origin" of a characteristic, a trait, or a disease (e.g. "genes are responsible for cancer or sexual orientation"). On the other hand, genetic responsibility can signify that individuals (should) deal with genetic information in a certain manner—namely, prudently and in accordance with moral imperatives. This meaning most often occurs in connection with reproductive decision making. The examples cited show, however, that genetic responsibility can apparently extend beyond the issue of reproduction to include the possible dissemination of genetic information to third parties. Should relatives be informed about genetic diseases or risks? Finally, to what extent are individuals themselves accountable for their illnesses, should they adjust their lifestyle to their genetic risks?[19]

The semantic complexes of causality/competence and guilt/duty do not, however, cover the entire spectrum of possible meanings. Two further aspects should be mentioned at least briefly. First, the discourse of genetic responsibility cannot be separated from calls for "responsible genetics." To the extent that more and more technological interventions based on the knowledge of cellular and molecular processes are possible, the problem of drawing legal and ethical boundaries becomes increasingly acute. This issue already emerged in the mid-1970s at the Asilomar conference, when guidelines for genetic research were drawn up at the instigation of scientists for the first time.[20] The appeal to responsibility has served since then to distinguish legitimate uses of genetic technologies from illegitimate ones.

The crucial point is of course to define criteria and conditions of a "responsible," that is to say the socially acceptable or ethically admissible employment of genetic technologies. A variety of actors, often with conflicting or even contradictory interests, address this issue. The demand for "responsible genetics" is promoted by biotech firms and the diagnostics industry, but also by watchdog groups and intellectuals critical of genetic technologies. For the Council for Responsible Genetics in the United States, this programmatic goal is already formulated in the organization's title. The discourse of responsibility also has great currency in patient organizations and self-help groups for genetic diseases. Thus, the tasks and aims for example of the Deutsche Huntington-Hilfe include, in addition to the improvement of the preventive and care situation of the patients and participation in scientific research on HD, ensuring "the ethically responsible application of the opportunities of genetic research" (Deutsche Huntington-Hilfe 2001: 57).

Second, it has to be noted that both semantic components—causality/competence and guilt/duty—do not simply exist in an additive relationship to one another. Instead, the discourse of genetic responsibility gains its contours through the specific tension generated by the confrontation of a scientific with a moral conception of responsibility. It has often been remarked that the results of genomic research threaten traditional concepts of personal responsibility and individual autonomy.[21] Contemporary biology, with its search for genetic (and neurobiological) determinants for a multitude of traits and modes of behavior, seems to subvert the substantial basis for responsible action: individual decision making and choice.

However, this assertion is quite debatable. What we observe today is not the negation of individual responsibility by reference to genetic dispositions and inborn traits. The discovery of genetic factors that influence and regulate the expression of diseases and personal traits does not result in a position that forecloses the responsibility of the subject; quite the contrary, the increasing genetic knowledge is the central point of reference for arguments about the expansion of moral duties. It engenders new modes and fields for responsible action. The next section will investigate this expansion of the concept of genetic responsibility in more detail.

Three Dimensions of Genetic Responsibility

The term "genetic responsibility" emerged in the mid-1970s in medical and bioethical literature. In 1974, a collection of essays appeared in the United States under the title *Genetic Responsibility: On Choosing Our Children's Genes*.[22] This book, edited by Mack Lipkin Jr. and Peter Rowley, discusses advances in medical genetics, the consequences of new reproductive technologies and the use of genetic screening. The problem of "genetic responsibility" arising from the new technological possibilities, however, is only addressed in an essay by Garrett Hardin, then a biologist at the University of Santa Barbara (California). In his contribution, titled "The Moral Threat of Personal Medicine," Hardin combines two lines of argumentation. On the one hand, he draws on evolutionary theory, claiming that medical interventions render the processes of natural selection inoperative and thus increase the "genetic burden" of coming generations. On the other hand, he stresses economic constraints in pointing out that the realization of a right to health or the provision of the best possible medical care in the context of genetic diseases results in enormous economic costs. Hardin suggests the concept of genetic responsibility as a necessary corrective to this trend:

> We must admit that if there is one thing a person is not responsible for, it is the genes that were passed on to him . . . We are not responsible as the recipients of errors. But should we not be responsible as the transmitters of errors? If there are some people in society who refuse to take such responsibility, who say No for whatever reason, refusing to inhibit their own breeding in spite of the fact they are passing on genes known to be undesirable genes, does not then the issue of responsibility arise in a very acute form? . . . Should individual freedom include the freedom to impose upon society costs that society does not want? . . . We must recognize that this is a finite world. The money we spend for one purpose, we cannot spend on another. (Hardin 1974: 88)

The problem of "genetic responsibility" is also addressed in another anthology published in the same year which deals with the ethical, social, and legal implications of genetic screening (Bergsma 1974). Sumner B. Twiss, in an article contribution titled "Ethical Issues in Genetic Screening: Models of Genetic Responsibility," argues that the spread of genetic technologies in pregnancy care and the screening of newborns and potential "carriers" has raised a significant problem not systematically addressed until now: the issue of "genetic responsibility" (1974: 227). Twiss pursues two goals in his text: "The predominant aim of this inquiry will be to provide an analytic and empiric account of emerging conceptions of genetic responsibility in order to explore their social and moral implications. A secondary goal is to lay some preliminary groundwork for proposing a minimal 'ethic of genetic responsibility.'" (Ibid.: 227–228)

In the course of his analysis, Twiss distinguishes between five models of genetic responsibility which are all based on the role of the parents. They concern possible "genetic duties" (ibid.: 234) of parents as regards the child, the family, society, humanity, and the tested individual's ethnicity. Twiss extensively discusses the obligations which in his opinion arise from these relationships in regard to the passing on of a "'serious genetic defect'" (ibid.. 235). The author is not concerned here with an abstract systematization or a heuristic exercise without empirical relevance; instead, he regards each form of "genetic responsibility" as an already existing social fact—a qualification, however, that applies to some forms more than others (ibid.: 233–244).

After this analytic differentiation, Twiss turns to the question of whether the use of genetic screening tends to become a moral obligation. The author maintains that there is a series of indications that this, in fact, applies to all five models of genetic responsibility. It can be observed, he writes, that individuals and family members are subject to social pressure to employ diagnostic technologies in prenatal care or to agree to their use. Twiss, however, is less interested in the existence of moral imperatives or social constraints in themselves than in the bioethical question of whether these are normatively justifiable. He weighs the arguments for and against limiting the freedom of parents as well as the duty of "responsible" reproduction and comes to the conclusion that two models of genetic responsibility can be justified: the responsibility of parents to their (unborn) child and that of parents to family members. Twiss defends a "minimal ethic of genetic responsibility," which he also terms an "acceptable view of genetic responsibility":

> Parents have the right to determine, according to their own wishes and notions of what is good for them, the genetic quality of their offspring, with the proviso that in certain circumstances they may have the duty to avoid bearing children with serious genetic defects if this is possible . . . Moreover, parents have a duty to permit relevant genetic information to be transmitted to relatives in the extended family, if this is medically/genetically indicated . . . The point here made was that genetic information impinging on the welfare of other family members is by its very nature not merely a matter of purely personal and private concern. (Ibid.: 258–260)[23]

Since the mid-1970s, the discourse of genetic responsibility has expanded significantly. The texts discussed here focused on the relationship of parents to their (future) children, deriving from this a potential duty to inform relatives about possible genetic risks. In the last few years, the development and use of postnatal genetic tests for numerous diseases have brought to light further facets of genetic responsibility, which will be examined in the following. According to a widely accepted assumption, genetic tests provide vital information which should "inform responsible decisions" in a variety of fields and situations (Kitcher 1996: 71).

In addition to *reproductive responsibility*, which is still called for (i.e. the concern for "healthy" descendents and the prevention of any transmission of "faulty" genes), two further semantic complexes have emerged that were absent from the earlier texts.[24] The first one is already mentioned in Twiss' article but is limited there to reproductive behavior: the *informational responsibility* to communicate potential genetic risks to (biological) family members. The new aspect is that not only is the issue of the possible transmission of "defective" genes to the next generation at the forefront, but so too are the direct health effects on existing family members who might be affected by genetic diseases in the future. Should relatives be informed of genetic risks so that they can opt for preventive and therapeutic options, to the extent that these are available? Where this is not possible, should they know about genetic risks in order to take "responsible" decisions regarding their family and life planning and so that they can inquire about diagnostic options?

The second aspect of genetic responsibility that has emerged since the 1970s refers not to the relationship to third persons but to an individual's health behavior and duties to him- or herself. *Self-responsibility* is called for not just concerning others, such as descendents and family members, but also in dealing with one's own genetic risks.[25] Genetic responsibility is specified in this case as the demand for genetic testing devices and for prevention schemes. In this perspective, only the knowledge of individual genetic risks allows a responsible lifestyle. To the extent that diseases are seen to have a (strong) genetic component, "mature" or "risk competent" health behavior requires specific knowledge of one's genetic risk profile, in addition to knowledge of general risk factors such as tobacco, alcohol, and lack of exercise.[26]

On the whole, three dimensions of genetic responsibility can be analytically distinguished: reproductive responsibility (prevention of transmitting genetic risks), informational responsibility (communication of genetic risks), and self-responsibility (monitoring of genetic risks). The discourse of genetic responsibility tends to replace fundamental rights by establishing relationships of duty to others and to oneself. It risks undermining essential legal guarantees and the freedom of choice. As an increasing number of women are now offered prenatal testing and predictive tests for more and more conditions, giving birth to a disabled child is often considered to be a matter of choice rather than fate. It appears to be more a blameworthy failure of surveillance and control than an unfortunate piece of bad luck.[27] The informational responsibility toward relatives, the duty to inform them of possible genetic risks, conflicts with the protection of the private sphere and the right not to know about genetic risks. The confidential relationship between physician and patient might be endangered by the obligation to disclose genetic data. Finally, the call for self-responsibility in dealing with genetic risks threatens to erode the individual right not to know. Seen from this perspective it is possibly irresponsible not to want to know—at least in cases when specific preventive and therapeutic options are available.

In the following section, I will trace the contours of "genetic responsibility" by examining relevant legal cases decided in the United States. These illustrate the creeping establishment of both informational responsibility and self-responsibility in using genetic knowledge and technological options available on the basis of this knowledge.[28]

INFORMATIONAL RESPONSIBILITY: COMMUNICATING GENETIC RISKS

Genetic information not only concerns the individual but may also indicate health risks for biological kin and future descendents. This specific quality of genetic data raises a number of important moral and legal issues. The question arises whether individuals should pass on genetic test results to potentially affected relatives (Andrews 1997; 2001: 43–55). The issue at stake here revolves around the conflict between the right to "genetic privacy" (Allen 1997; Sommerville and English 1999; Meyer 2001: 208–216) and the responsibility to inform relatives of health risks. Medical practice is furthermore concerned with the tension between the principle of the confidentiality of the doctor–patient relationship and the duty to warn patients that they are at risk for a genetic disease. Under what conditions are physicians legally obliged to disclose genetic information that is medically relevant to potentially affected relatives? When do they have to respect the "genetic privacy" of their patients (Orentlicher 1997; Pergament 1997)?

There are different regulatory frameworks to deal with the problem of confidentiality in the context of genetic information. Although in most countries regulations protect the privacy of genetic information, there remains considerable legal uncertainty when stipulating situations in which such information may be disclosed without liability. In France, any direct transmission of genetic information to other persons or institutions is forbidden, whereas in the United Kingdom, in Germany, and in the United States, as in many other countries, the right of confidentiality is upheld in principle but it may be restricted under certain conditions, as, for example, when serious health risks can be avoided. The *American Society of Human Genetics* thus states:

> Disclosure should be permissible where attempts to encourage disclosure on the part of the patient have failed; where the harm is highly likely to occur and is serious and foreseeable; where the at-risk relative(s) is identifiable; and where either the disease is preventable/treatable or medically accepted standards indicate that early monitoring will reduce the genetic risk. (American Society of Human Genetics 1998: 474, 480–482; British Medical Association 1998: 68–73, 146–150; cf. the overview in: Parker and Lucassen 2002: 1686)[29]

In Germany, too, the principle of medical confidentiality is subject to continual weighing of the advantages and disadvantages.[30] The *Guidelines on Predictive Genetic Diagnostics* issued by the German Medical Association draw attention to the "special significance" of the preservation of medical confidentiality and the observance of data protection regulations as regards genetic knowledge, as it is "of a personal and predominantly intimate nature" (Bundesärztekammer 2003: A 1303). On the other hand, knowledge provided by predictive tests also creates "special difficulties" (ibid.) because it allows conclusions about the genetic constitution of family members to be drawn who were not themselves involved in the medical examination. Acknowledging this tension, the German Medical Association proposes "a consideration according to the degree of affliction and according to the possibilities of secrecy." If the patient does "not agree," the physician is allowed, "in case of a specific situation of danger, to pass on the confidential information to third persons" (Ibid.).[31]

German medical law provides for only three admissible exceptions to the obligation of medical confidentiality: "The disclosure of patient data is only admissible when it is authorized either by the patient's consent, by a legal regulation, or by a justifiable reason—in practice the most problematic case." (Henn 2002: 344) Such a justifiable reason was deemed applicable by the Higher Regional Court in Frankfurt am Main in a judgment in 1999. This case concerned the proper balance between two legally protected interests: on the one hand, trust in the family doctor, and on the other, the life and health of an acutely threatened person. A man infected with HIV had forbidden his family doctor to inform his female partner of his infection. When the woman, a patient of the same physician, discovered after her partner's death that he had had AIDS and then tested positive for HIV herself, she sued the physician for damages. The Higher Regional Court ruled in favor of the claimant and decided that the family doctor should have informed the woman. The Higher Regional Court judged that the physician not only had the right but also the duty to inform the woman. The court stated that the legally protected interest of the trust of the infected patient in his attending physician had less weight than the legally protected interests of "life" and "health" of the acutely threatened woman. According to the Higher Regional Court, the AIDS patient had exhibited a lack of a "sense of responsibility." Thus, it confirmed the legal obligation of the physician to inform the claimant about the existing danger, even against the declared will of her partner (ibid.: 345).

Both the judgment of the Higher Regional Court and the explicit exception from the principle of medical confidentiality referred in this specific case to an infectious disease. This judgment therefore gives no unambiguous indication concerning the outcome of a case involving genetic risks. Nevertheless, it is reasonable to assume that the same preventive logic would most certainly be cited. Genetic counselors have pointed out that it is possible to compare genetic diseases to infectious diseases in this respect. Wolfram Henn, for example, cites a (genuine) case of hereditary intestinal cancer to illustrate the problem of medical confidentiality as regards relatives. In this specific case, the patient did not want to inform relatives possibly affected by the same disease due to inter-familial conflicts, and forbade the physician from making contact with them. According to Henn, a conflict situation is created here whose legal consideration entails a similar judgment to that reached by the Higher Regional Court in Frankfurt.

> This case concerns a weighing of legally protected interests, which strongly recalls . . . the case of the HIV infection that was kept secret. Here, too, the obligation of care for the risk person precedes the obligation of medical confidentiality, so that proactivity on the part of the physician is justified as *ultima ratio*. (2002: 348)

Henn therefore proposes, in accordance with the code of conduct issued by the German Medical Association, that in genetic diagnoses which allow conclusions regarding potential genetic risks of relatives to be drawn, a right, but not an obligation, of the physician to violate the imperative of medical confidentiality should be recognized.

Beyond considerations and assessments put forward by genetic counselors on this issue, the analogy between infectious and genetic diseases has already shaped judgments of higher courts and has served to confirm a legal obligation to inform relatives. In two cases, high courts in the United States ruled that doctors are obliged to warn children of a patient that may be at risk genetically of acquiring a disease from their parent. The first case was decided in 1995 by the Florida Supreme Court (*Pate v. Threlkel*, 661 So. 2d 278 [1995]).[32] This case was brought by Heidi Pate, whose mother was diagnosed with medullary thyroid carcinoma in 1987. After Ms. Pate learned that she had the same disease in 1990, she and her husband sued her mother's physician. The lawsuit alleged that the physician was under a duty to tell the mother of the importance of testing her children for medullary thyroid carcinoma because of its hereditary character. The plaintiff claimed that had the physician warned her, she would have been tested, and the mother would have taken preventive action before the daughter developed the disease. The trial court dismissed the plaintiff's claim. It ruled that because Heidi Pate was the patient's daughter, there was no professional relationship with the physician. As a consequence, no substantial ground for bringing a malpractice suit existed. However, because this was the first time such an issue had arisen in Florida, the case was transferred to the Florida Supreme Court.

The Florida Supreme Court ruled against the physician. It concluded that the physician's alleged duty to warn extended to the children of the patient even though the children were not his patients: "We conclude that when the prevailing standard of care creates a new duty that is obviously for the benefit of certain identified third parties and the physician knows of the existence of those third parties, then the physician's duty runs to those third parties." (*Pate v. Threlkel* 1995: 282) The Supreme Court then discussed how the duty could be discharged. For practical reasons and in line with professional principles, it ruled that the duty does not require the physician to warn the patient's children of the disease but would be satisfied by warning the patient about the possible health risks for his or her relatives (Petrila 2001: 407–408; Deftos 1998: 964; American Society of Human Genetics 1998: 480).[33]

One year later, a similar case was decided by an appellate court in New Jersey that further extended the professional duties of the physician (*Safer v. Pack*, 677 A.2d 188 [App. Div. NJ, 1996]). In this case, the father of the plaintiff, Donna Safer, was treated for cancer in the 1950s and 1960s. He was hospitalized in the beginning of the 1960s with colon cancer and died in 1964 when she was ten years old. In 1990, Donna Safer was diagnosed with colon cancer herself. After obtaining her father's medical records, she filed suit in 1992 against the estate of her father's doctor (who had died in 1969). She alleged that her father's physician knew or should have known of the hereditary nature of the disease, and she saw a violation of a duty to warn her of the risk to her health while her father was treated. According to Safer, a timely warning would have enabled her to take preventive measures to reduce her risk (Petrila 2001: 408–410; Clayton 2003: 566–567).

Although the trial court denied the plaintiff's motion of judgment, the appellate court came to a different conclusion. Like the Florida Supreme Court, the New Jersey court held that the physician's duty extended well beyond the patient to include his children. The court confirmed that there is a legal duty to warn those at risk of avoidable harm from genetically transmissible conditions. A very important reason for this judgment was the fact that the court treated genetic risks just like any other type of medical risk, thereby assimilating genetic risks to infection risks. According to the court, there was "no essential difference" between "the type of genetic threat at issue here and the menace of infection, contagion or threat of physical harm . . . The individual at risk is easily identified, and substantial future harm may be averted or minimized by a timely and effective warning" (*Safer v. Pack* 1996: 1192, cited by Petrila 2001: 409).[34]

In this perspective, the transmission of genetic risks by parents appears to be quite similar to the problem of a possible infection, and as there is a duty to warn relatives in the case of contagious diseases the same principle applies to genetic risks. However, as the American Society of Human Genetics correctly remarked, the contagious disease model "is not an ideal paradigm for the disclosure of genetic information" (1998: 477).[35] There are several substantial differences between genetic risks and infection risks. The first concerns the etiology of the disease. Whereas genetic conditions are transmitted "vertically through succeeding generations . . . contagious disease is generally transmitted horizontally . . . and its impact on others occurs through some form of contact." In contrast to infection risks, genetic risks cannot be separated from the person herself or himself; they are not temporary and accidental, but an integral part of the person's physical constitution.[36] The second difference concerns the type of intervention. Contagious disease is "controlled by isolation of infected people, by avoidance of whatever contact causes infection, or by cure. Genetic conditions, on the other hand, are controlled not only through prevention or palliative treatment but also through reproductive decisions and choices." (Ibid.) The "identifiable victim" (cf. *Safer v. Pack* 1996: 1192) of genetic risks not only refers to the already living, but also—or more precisely, above all—to future generations and individuals not yet born. There is a third difference that points to the consequences of the analogy between infection risks and genetic risks: warning relatives about genetic risks will not prevent them from having the gene (Andrews 1997: 268–269; see also Deftos 1998: 966; Petrila 2001: 412; Clarke 2007: 23).[37]

In spite of these profound differences between genetic risks and infection risks as regards the pathogenesis and forms of intervention, the analogy can look back on a long historical tradition. Diane B. Paul (1998b) has drawn attention to the fact that the scientific discussion of inheritance and hereditary diseases has been conducted in bacteriological terms since the beginning of the 20th century. Genes were often described as infectious germs, and "genetic defects" provoked fears similar to those surrounding a hidden plague. Talk of "carriers" also referred to the danger of transmission following the model of infectious diseases (see also Gaudillière and Löwy 2001).[38] According to Peter Conrad, the current success of genetic concepts of disease can be traced to the fact that in their essential aspects they rest on the bacteriological paradigm. Both focus on the interior body and on clearly identifiable "carriers" in the search for the causes of disease. Although the vision of a "molecular medicine" invokes an etiological paradigm shift, at the same time it resorts to established models of explanation and traditional "thought styles" (Fleck 1981):

> [T] he close fit between germ theory and gene theory is one of the chief reasons that genetic explanations have been so readily accepted in medicine and the popular discourse. At least on the level of assumptions and structure, gene theory does not challenge common conceptions of aetiology but rather shifts its focus. In this sense at least, genetics is a

complementary rather than a challenging paradigm in medicine. (Conrad 1999: 232; see also Steinberg 1996)

The epistemic rapprochement of genetic risk to other forms of medical risk allows for a normative extension of genetic responsibility. The two court decisions institutionalize a legal duty of the physician to warn if he or she knows or should have known that the patient's children are exposed to a genetic risk that is related to the disease diagnosed in the patient. In both cases, it was held that there are legal obligations that go beyond the concrete relationship between the physician and his or her patients toward the patient's children. This development is not confined to Germany and the United States; nor does it exclusively affect the legal system. Rather, it is embedded in more general transformations. In medicine and bioethics, a "proactive approach" (Suthers et al. 2005: 668) according to which geneticists should actively notify family members of patients of possible genetic risks is gaining more and more ground (see Newson 2006; Minkoff and Ecker 2008).

This creeping tendency to establish a duty to warn in the context of genetic diseases results in some serious problems.[39] First, the question arises of whether the physician's obligation to inform does not in fact undermine the right of relatives not to know. Family members, whom the physician is now legally obliged to warn, might not want to know whether they carry an increased genetic risk. How can the right not to know be maintained at all under the conditions of a legal obligation of the physician to inform the relatives? Second, one must take into account that the risk calculation and the predictive value can be very different depending on the type of disease and the pattern of inheritance. Thus, the genetic risk for those possibly affected might be assessed as high in some cases (e.g. for Huntington's disease) and low in other cases (e.g. for hereditary hemochromatosis). Is the legal obligation to inform a patient's children of possible genetic risks also operative in such cases where the link between a (presumed) genetic variation and the emergence of a specific disease is fairly vague? To the extent that genetic factors are made responsible for an increasing number of diseases, the duty to give a warning would in fact be tremendously expanded. To what extent should the courts' judgments be applied to diagnoses of mental disorders? Are relatives also to be warned when genetic components could play a role in the manifestation and development of depression or schizophrenia?

Third, there are some very practical problems that need to be addressed. How can it be ensured that the patients informed by their physician communicate their genetic risk status to their children appropriately? One must bear in mind that affected patients might not want to tell their children about their possible genetic risks, perhaps because they consider them medically irrelevant or because they feel morally guilty about transmitting genetic risks to their sons and daughters. The dissemination of this information could have significant effects on family relationships and lead to tensions between family members.[40] Fourth, there is the question of the extent and duration of the physician's legal obligation. Which persons and timeframes does it apply to? What if the children are born after the diagnosis of the disease for the parent? Similarly, should a physician have a duty to inform his or her patient of new advances in genetic knowledge (even after the termination of the patient–physician relationship)?

Jorgen Husted has pointed out that the institutionalized obligation to warn family members has in fact abolished the right not to know: "As it seems, the alleged right to know does not comprise a right not to know . . . A wholly new situation has been created—the irreversible loss of genetic informational innocence is a fait accompli." (1997: 56) During medical treatment, the patient has given his or her informed consent to a genetic test and to the revelation of the test's results, but this is not the case for a family member who is informed afterward. The question arises, however, of whether one should not presuppose a tacit consent, a vital interest in being informed of possible genetic risks. Is the passing on of genetic information to family members not just as much proof of responsible behavior, as it can be expected that the receivers of the information will be interested in this information whatever the circumstances because it provides the basis for prudent and future-oriented health behavior? This assumption of a pre-established harmony of differing responsibilities is the view taken by the renowned Nuffield Council on Bioethics in their justification of the obligation to warn relatives of genetic risks:

> As a starting point, we adopt the view that a person acting responsibly would normally wish to communicate important genetic information to other family members who may have an interest in that information, and that a responsible person would normally wish to receive that information, particularly where it may have a bearing on decisions which he or she may be called upon to take in the future. (1993: 49)

Seen from this perspective, there is no conflict of norms. Quite on the contrary: to transmit genetic information to relatives can be seen as an important indicator for a well-ordered and functioning family life, as a test for responsible behavior and care for each other, and as a symbol of superior moral quality.[41] Thus, the discourse of genetic responsibility allows potentially diverging interests to be reconciled with one another. In this perspective, the patient who acts "responsibly" by passing on genetic information to relatives can count on a "rational" subject, who "normally" wishes to use the information received for health management.

The principle of informed consent, which originally served as a critique directed against medical paternalism, is now transformed into its opposite. The calls for "informed consent" and "patient autonomy" were at one time used against the physician's exclusive authority to interpret and treat. This challenged the view that the physician, according to his or her assessment, could with or without the patient's permission suggest a certain medical procedure considered by the physician to be in the patient's best interest. The call for the patient's "informed consent" thus served to reduce medical paternalism. As Husted makes clear, this principle has apparently been relegated once more to the background within the context of genetic responsibility, and has given way to a resurgence of paternalist practices:

> Thus the unsolicited disclosure, whether by the relative following the doctor's strong suggestion or by the doctor acting independently, seems to be a clear cut case of strong medical paternalism—acting solely from the medical point of view the decision "To know or not to know?" is taken out of the hands of the unsuspecting individual, for her or his own good of course. (1997: 57)

The conflict of interest between an obligation to inform family members and their right not to know is reproduced in the contradictory practical demands on the attending physician who should take care that patients deal responsibly with genetic information. Although physicians are to avoid every form of influencing and directive behavior, they would appear to be sufficiently qualified for the moral task of informing patients about the life style changes and the informational duties which are to follow from genetic knowledge. Medical expertise, thus, steps down in favor of moral competence (ibid.: 68).

SELF-RESPONSIBILITY: MONITORING GENETIC RISKS

In addition to the duty to inform others of possible genetic risks, genetic responsibility is increasingly being addressed to the individual as responsibility toward the self. In this sense, it refers to the prudent management of genetic risks through informed lifestyle choices that are based on genetic information. Genetic responsibility here means an active demand for genetic information and an interest in genetic testing options. This scenario was well described by a hypothetical case that was presented as long ago as 1999 by Francis Collins, then director of the National Human Genome Research Institute of the United States:

> John, a 23-year-old college graduate, is referred to his physician because a serum cholesterol level of 255 mg per deciliter was detected in the course of a medical examination required for employment . . . To obtain more precise information about his risks of contracting coronary artery disease and other illnesses in the future, John agrees to consider a battery of genetic tests that are available in 2010. After working through an interactive computer program that explains the benefits and risks of such tests,

> John agrees . . . to undergo 15 genetic tests that provide risk information for illnesses for which preventive strategies are available . . . A cheek-swab DNA specimen is sent off for testing, and the results are returned in one week. John's subsequent counseling session with the physician and a genetic nurse specialist focuses on the conditions for which his risk differs substantially (by a factor of more than two) from that of the general population . . . John is pleased to learn that genetic testing does not always give bad news—his risks of contracting prostate cancer and Alzheimer's disease are reduced, because he carries low-risk variants of the several genes known in 2010 to contribute to these illnesses. But John is sobered by the evidence of his increased risks of contracting coronary artery disease, colon cancer, and lung cancer. Confronted with the reality of his own genetic data, he arrives at that crucial "teachable moment" when a lifelong change in health-related behavior . . . is possible . . . His risk of colon cancer can be addressed by beginning a program of annual colonoscopy at the age of 45, which in his situation is a very cost-effective way to avoid colon cancer. His substantial risk of contracting lung cancer provides the key motivation for him to join a support group of persons at genetically high risk for serious complications of smoking, and he successfully kicks the habit. (1999: 34–35)[42]

The fictive example clearly indicates that the proliferation of genetic knowledge and testing devices does not abolish individual responsibility and choice; it rather produces, if seemingly paradoxically, a new form of autonomous subjectivity. The individual is conceived not as a passive recipient of medical advice, but as an active seeker of information and consumer of genetic testing devices and healthcare services (Petersen and Bunton 2002). The social significance of genome analysis and genetic diagnostics lies less in the deterministic information that they seem to offer, and more in the "reflexive" relationship that they seek to generate between an individual risk profile and social requirements.[43]

Bioethicist Hans-Martin Sass therefore calls for an "ethos of duty" in handling genetic information:

> Leisure time behavior, place of work, or genetic predisposition, or a mixture of all three factors determine[s] the respective individual risks to my health . . . Some can be eliminated, others reduced, or the stage at which they become acute delayed. The patient becomes the partner in preventing or delaying major health risks. The doctor's ethics under the Hippocratic Oath, characterized by care and outer-determined support, will in future be complemented by a self-determined and self-responsible ethics of the patient and citizen in healthcare. (1994: 343)[44]

In this perspective, genetic factors are not regarded as the opponent of autonomous behavior but, on the contrary, as its object and material.

Individual decision making must incorporate not only lifestyle factors such as smoking or alcohol, but also extends to the "responsible" handling of genetic information. This argument is confirmed by statements from key representatives of the medical profession and bioethicists.

The former chairman of the Deutsche Gesellschaft für Humangenetik [German Society for Human Genetics], Holger Höhn, expressed this trend toward genetic self-responsibility and the role of human genetics in the following words:

> What characterizes human genetics . . . is the prediction, the forecasting of a phenotype, of the course of a disease, of the probability of falling ill. No other discipline currently fulfils the requirements for a methodically flawless and broadly thought-out prediction of the biological life course of an unborn or born person. No other discipline is capable of so extensively informing patients of the scope and implications of this prediction, so that they are able to take self-responsible decisions concerning their life, take their life into their own hands. (1997: 173)

For the former president of the Deutsche Forschungsgemeinschaft [German Research Foundation], Hubert Markl, individual values and moral concerns are pertinent in the context of genetic knowledge, because compulsory tests and a eugenics imposed by the state are today forbidden:

> Surely no-one who does not want to know about his hereditary disposition can be forced to take a genetic test. But if someone has voluntarily—or compelled by illness—acquired knowledge of a genetic disposition for disease, then *individuals* are confronted with the moral issue of whether they should not modify their lifestyle so that the onset of the disease is delayed or prevented. (1989: 62; emphasis in original)

However, the autonomy in question relies on a rather specific and liberal conception of morality. It is limited insofar as the individual is considered an abstract subject free from material restrictions, cultural values, and social bonds. He or she is forced to choose from an array of pre-established options and to take responsibility for the consequences of his or her choices. Only those actors who accept health as a superior social and moral value and who regard biomedical and scientific expertise as essential for everyday life and personal decision making qualify as rational or responsible subjects (Beeson and Doksum 2001; Callon and Rabeharisoa 2004). This genetic enlightenment (or "literacy") entails a precise notion of "Mündigkeit" [maturity] which is linked to possessing adequate medical information and to the knowledge of one's genetic risks (Sass 2003). It contributes to constituting a "homo geneticus" (Gaudillière 1995: 35; Novas and Rose 2000) who submits to practices of self-control and personal management of the body, which comprises an embodiment of risk technologies and genetic knowledge.

The reference to personal responsibility and self-determination only makes sense if the individual is more than a victim or prisoner of her or his genetic material. If there were indeed a monocausal relationship between genotype and phenotype in the sense of genetic determinism, then it would be much harder to uphold the appeal to individual autonomy. By contrast, the construction of "at-risk" individuals, families, and pregnancies makes it easier to moralize about deviant behavior and to assign guilt and responsibility (Douglas 1990). Paradoxically, it is exactly the invitation to engage in self-determination and the imperative of a "genetic responsibility" that render individuals more and more dependent on medico-scientific authorities and their information. The right to health is realized in the form of a duty to procure medically relevant information. In this perspective, only people who act responsibly are those who draw the correct, that is to say, risk minimizing and future oriented, conclusions from this range of information. The jurist Reinhard Damm has observed an "inverted ethics of responsibility," which attempts to attribute the costs and consequences of genetic diseases or risk diagnoses to those affected:

> While until now solidarity and non-discrimination of the ill/patients had the highest priority in medical, labor or insurance law, today the emphasis has shifted in the direction of the burden of responsibility of the genetically afflicted in familial, labor or insurance relations. The responsibility can develop social pressure with regard to compulsory gene tests as well as to the disclosure of genetic data. (1999: 448)

Beyond bioethical discussions and recommendations, there are tendencies within legal policy which invert the attribution of the burdens of responsibility. This is demonstrated by a Supreme Court decision in the United States that determined the limits of the non-discrimination law in the United States in a judgment in 2002 curbing the rights of employees (*Chevron USA Inc. v. Echazabal*, Case No. 00–1406, US. Supreme Court, Washington, DC [2002]).[45] The plaintiff, Mario Echazabal, had worked at a Chevron oil refinery in El Segundo, California since 1972, as a laborer, helper, and pipe fitter for various contractors, mainly in the coker unit. In 1992, he applied to work directly for Chevron at the refinery's coker unit. Chevron determined that Echazabal was qualified for the job and offered to hire him contingent on the results of a physical examination. The company doctor, however, declared Echazabal unfit for the job because blood tests showed liver abnormality. During the examination, a liver function assay was used that resulted in the identification of a biological marker that, according to the employer, disposed Echazabal to liver impairment. As a consequence, he would face further damage if he experienced chemical exposures characteristic of refinery work. Nonetheless, Echazabal was permitted to continue working at the company as an employee of Chevron's contractor. He sought treatment and was ultimately diagnosed with hepatitis C that has remained asymptomatic since then. In 1995, he again applied for a job at Chevron, and again his demand was turned down after a medical examination. This time, the company directed the contractor who employed Echazabal to remove him from that position, which it did in 1996. This action was taken even though Echazabal's liver condition never caused injury or accident to himself or anyone else at the refinery. In 1997 he filed suit, charging that Chevron's decision violated the Americans with Disabilities Act (ADA). He also presented testimony by two medical experts on liver disease that working in that factory would not put him at any greater risk than any other employee. Chevron defended itself under a regulation of the Equal Employment Opportunity Commission (EEOC)[46] permitting the defense that a worker's disability on the job would pose a "direct threat" (*Chevron v. Echazabal* 2002: 1–3; National Council on Disability [NCD] 2003: 5–6).

After a district court granted summary judgment for Chevron and a circuit court reversed this decision, the case was presented to the U.S. Supreme Court. The Supreme Court addressed issues central to the ADA in its decision. At the heart of the ruling is the court's interpretation of the "direct threat" provision of the ADA, because an individual may be refused employment if a direct threat can be established. The previous ruling in this case by a circuit court held that the direct threat defense was not available to Chevron because Echazabal only presented a risk to himself. According to the court, a "direct threat" only applies when the individual's condition poses a direct threat to others. In this perspective, health is conceived as a "discretionary right in which the individual may choose to assume certain risks so long as they do not have the potential to harm others" (Lomax 2002: A505). This reasoning was rejected and reversed by the Supreme Court. The court held that individuals who pose a risk exclusively to themselves may be excluded from a job as long as the employer relies on "reasonable medical judgment" (*Chevron v. Echazabal* 2002: 12). The company thereby sought to "protect" an individual such as Echazabal from himself.

By presenting evidence from a biological marker, Chevron relied on a medical opinion that is based on probabilities. The same is true for genetic susceptibility testing.[47] Because the ADA is often cited as offering individuals protection from genetic discrimination, one might ask the (speculative) question of whether the decision would have been different in this case. According to the Supreme Court, it is acceptable to exclude from a job individuals who pose a risk exclusively to themselves as long as the employer relies "on the most current medical knowledge and/or the best available objective evidence" (*Chevron v. Echazabal* 2002: 12). This condition might be fulfilled by genetic susceptibility testing that is increasingly acquiring medical and scientific credibility. As a consequence, it is conceivable that future workplace exclusion might be based on the diagnosis of genetic susceptibilities. The Supreme Court decision might be important for the "asymptomatic ill" (Billings et al. 1992), as the National Council on Disability (NCD), an independent federal agency that was instrumental in creating the legislative record that Congress considered in enacting the ADA, predicts: "With advances in medical technology, including genetic screening, there also is the potential for excluding large numbers of pre-symptomatic individuals (i.e. 'the healthy ill') on the basis of potential health or safety risks to themselves in the future." (2003: 11; see also Lomax 2002; Rouvroy 2008: 59–62)[48]

The NCD, which was decisively involved in the drawing up of the ADA, sees in this judgment the inversion of the goals once connected with non-discrimination law.

> Congress acknowledged in the ADA that discrimination takes many forms, including paternalism and stereotyping. Perhaps the most long-standing and insidious aspect of this type of discrimination is the assumption that people with disabilities are not competent to make informed, wise, or safe life choices. (2003: 9)

The court decision will make it much easier for employers in the United States to exclude workers by referring to the possibility of a "direct threat" that originates in a higher genetic susceptibility (ibid.: 15–16; see also Marshall 1999; Geppert and Roberts 2005).

Apart from its effect on disability rights, the *Chevron v. Echazabal* decision may have important implications for environmental health research (Lomax 2002: A504). Although research in this field has traditionally concentrated on identifying external risk factors that pose health problems to employees (exposition), more and more scientific emphasis is now being placed on recognizing internal risks or personal susceptibilities that are based on the genetic make-up of individuals (disposition). In reviewing a report the Genomics Task Force for the U.S. Environmental Protection Agency (EPA) submitted in 2004, Susanne Bauer observes that genomic tools are granted a central role in determining individual and collective susceptibility profiles and the intensity and extent of clean-up practices. This new focus indicates a shift in the concepts and operational principles of environmental and occupational health that increasingly embrace differential models of protecting individuals and collectives against pollutants—with serious consequences for (environmental) equity and justice:

> Supplementing biomarkers of exposure by markers of genetic susceptibility tends to further frame the effects of exposure in terms of individual biology and, potentially, individual responsibility. It is the mobilisation into individual risk-management that can be seen as part of an emerging "biological citizenship", as society, environmental policies and the health care system appear reconfigured according to

> bioscientific categories. (Bauer 2008: 10; 2009: 127–129; see also Hoffmann 1999; Hennen, Petermann, and Sauter 2001: 103–116)

As a consequence the "old" risk logic is modified and supplemented. Scientific interest not only focuses on dangerous working conditions or toxic substances in the labor process but also on "supersensitive workers" (Daniels 2003: 548) or susceptible employees who are less resistant than others to environmental risk factors and health-threatening conditions of work. Similarly, a shift in the modes of prevention and points of intervention is observable. "Primary prevention in the sense of protection from environmental hazards is not at the center of efforts, but secondary prevention by singling and filtering out 'susceptible' and 'oversensitive' persons by means of genetic screening dragnets." (Klees 1992: 227; 1990)

The shift from "exposition" to "disposition" is illustrated by a further case from the United States. In 2001, it came to light that Burlington Northern Santa Fe Railway Company, a U.S. railway company with more than 40,000 employees, had begun obtaining blood for genetic testing from employees. The background is that 125 workers sued the railway company for damages because they suffer from the so-called carpal tunnel syndrome. The disease causes pain and numbness in the hands and arms, and is usually brought on by continually repeated movements. A genetic component for the condition is also suspected. A genetic test can establish whether or not there is a mutation which supposedly raises the risk of the disorder. The company did not explain why it carried out the test; it can be assumed, however, that it intended to refuse to pay damages by stressing the causal role of genetic factors for the disease. When the administration of the tests became public, the EEOC sued Burlington Northern Santa Fe Railway. A short time later, the railway company paid a large sum to the employees and the lawsuit was dropped. The chairwoman of the EEOC, Ida L. Castro, justified the lawsuit against the company by saying that

> the only possible explanation is that the railway company wanted to refuse to pay damages for a recognized job disease. The management's argument would have been: Who knows, maybe one or another employee would have had the carpal tunnel syndrome anyway due to that employee's genetic disposition. That is why we had to intervene. (Lehming 2001; Clayton 2003: 563; Miller 2005)

The regime of "genetic self-responsibility" can paradoxically, as in the Echazabal case, lead to a situation where the interests and preferences of those affected no longer play any role.[49] Autonomous decisions regarding one's own health risks are replaced by the principle that the company knows best what is good for its workers. If this trend spreads to other social areas, then the right not to know in social practice will soon yield to the duty to manage genetic risks. The discourse of genetic responsibility shows that the "genetic police" not only operates via prohibition, coercion, and exclusion, but functions above all as a "genetic civilisation strategy" (Ten Have 1997: 92), which promotes normative concepts and social judgments and materializes itself in institutional arrangements and legal regulations. This strategy suppresses and sanctions deviations from the norm, and also produces concepts of genetic normality—which are visible in social demands for a "responsible," "mature," and "rational" management of genetic risks: "[W]hen genetic information is likely to make a significant difference in my decisions and when the relevant information is obtainable with reasonable effort, I have no right to remain ignorant. From the recognition of my own autonomy, I have a duty to be informed." (Rhodes 1998: 18)

6 Conclusion
Pitfalls of Criticism

The present study has highlighted several important weaknesses and problems in the current debate on genetic discrimination. This last chapter summarizes the conceptual and empirical findings of the investigation and points to some serious shortcomings in the normative assessment and political evaluation of forms of exclusion, stigmatization, or disrespect based on genetic information. It calls for a more comprehensive concept of genetic discrimination which breaks with the idea of genetic normality that informs many non-discriminatory policies.

At a *conceptual* level, the study shows the limits of a case-oriented and person-centered definition of discrimination, which has to be enlarged in two respects. First, it has proved helpful to distinguish between direct and indirect forms of genetic discrimination. This differentiation makes it possible to examine social stereotypes and prejudicial structures which affect the individuals involved and constrain their scope for decision making and action. Whereas direct genetic discrimination describes the manner in which people with genetic characteristics are individually treated, indirect discrimination refers to prejudicial structures and forms of disrespect addressed to all members of society (cf. chapters 3 and 4). The widening of the analytic focus to include mechanisms of indirect discrimination corrects a weakness of previous studies, which have concentrated exclusively on coercive measures and asymmetrical decisionmaking processes. The emphasis on "negative" patterns of litigation has thus excluded the question of the extent to which additional options and symmetrical situations produce discriminating effects. On the basis of the definition of indirect discrimination proposed here, however, it is possible to examine how growing genetic knowledge functions as a moral technology, not merely prohibiting and excluding but also encouraging and promoting lifestyle patterns and forms of conduct by referring to the notion of "genetic responsibility." This concept, unknown until recently, engenders new moral duties and institutional expectations (cf. Chapter 5).

The second proposal concerns the distinction between interactional and structural discrimination. In addition to practices of genetic discrimination by organizational actors such as insurers, employers, and so on, forms of disrespect, stigmatization, and exclusion in social interactions also need to be included in the analysis. Interactional discrimination manifests itself in intentional forms of action but also in stereotypes, prejudicial structures, and patterns of interpretation which can lead to discriminatory practices even without the conscious intention of discrimination. In contrast, structural discrimination concerns entrenched and permanent, often formalized and explicitly regulated organizational practices. An example of this is the denial of insurance coverage due to calculations of genetic risk, whereas disrespect from individuals or groups (e.g. in families or from neighbors) on the basis of genetic data exemplifies interactional discrimination. However, the two levels of discrimination can only be distinguished from one another analytically; empirically, they do not mutually exclude one another at all but, on the contrary, intertwine and overlap regularly in social reality (cf. Chapter 4).

In contrast to indirect mechanisms of genetic discrimination, which function more through the mobilizing of normative expectations and moral judgments, structural discrimination is characterized by a higher density of legal regulation and a more comprehensive degree of formalization. Whereas indirect mechanisms might appeal, for example, to the moral responsibility of future parents to ensure that their child is healthy in order to promote the use of genetic testing during prenatal care, the "wrongful life" jurisdiction goes beyond this in that it legally treats a disabled child as damage to be compensated for.[1] In order to develop a feasible and clearly developed definition of genetic discrimination, however, an internal distinction between direct/indirect discrimination, on the one hand, and interactional/structural discrimination, on the other, does not suffice; further research will have to go beyond this and examine in what manner "genetic discrimination" can be differentiated from related concepts such as "stigmatization" (Sayce 1998; Link and Phelan 2001), "geneticization" (Lippman 1991), or "geneticism" (Wolf 1995) (cf. Bioethics Research Committee 2004).

This issue leads to the *empirical* findings of this study. Two points deserve special attention in this context. First, the study provides evidence that the principle of distinguishing between phenotypic and genotypic discrimination that has guided empirical studies of genetic discrimination in the past has to be questioned, for various reasons. To start with, this juxtaposition does not allow for the investigation of how forms of genetic discrimination are linked with sexist and racist practices and mutually reinforce one another (cf. Chapter 3). Furthermore, a strict demarcation between genotype and phenotype may not be very helpful in grasping the fears connected with genetic illness. On the basis of the experiences of exclusion and stigmatization reported by those affected by Huntington's disease, few substantial arguments could be found for a systematic differentiation into two isolated and distinct forms of discrimination and groups of affected persons (asymptomatic "risk persons" versus symptomatic patients). Future research should therefore determine if and to what extent *genetic* discrimination represents an analytically distinct, singular form of discrimination, essentially different from discrimination against the disabled or the chronically ill. However, one must keep in mind that a comprehensive definition of genetic discrimination which expands the circle of those potentially affected to the (chronically) ill and disabled ultimately becomes counterproductive, as the uniqueness of the different treatment of asymptomatic "risk persons" may be lost (Chapter 4).

Second, the study has presented empirical evidence that gives rise to a different account of the public assessment and legal regulation of genetic risks. On the one hand, it is evident that the media and political debates assume a special epistemological and normative status for genetic factors, and that this also shapes non-discrimination legislation. Here we regularly find essentialist and deterministic accounts of the interaction between phenotype and genotype. In this perspective, individuals affected by genetic risks should enjoy particular legal protection, because they are supposedly not responsible for their genetic constitution and genes are conceived of as the basis of human identity (Chapter 3). On the other hand, the bioethical and medical literature and legal decisions demonstrate a creeping tendency to negate substantial differences between genetic and other medical risks, assimilating the one increasingly to other. From this perspective, genetic knowledge is not different from conventional medical knowledge providing information for "rational" health management. The contours of a "genetic responsibility" that imposes new moral and institutional obligations on individual decisions are becoming increasingly evident. The search for genetic causes of disease does not lead to the negation or limitation of the subject's responsibility, but instead expands existing duties and establishes new areas of responsibility (Chapter 5).

This brings us to another important finding of this study, which draws critical attention to the theoretical weaknesses and normative ambivalences of the non-discrimination debate. Current legal regulations determining the foundation and limits of the use of genetic data suffer from three central problems. They are based on the erroneous assumption that genetic and non-genetic information, testing procedures, and diseases can be clearly separated from one another; they resort to the essentialist notion that genes define the essence of humanity; and they ultimately lead to inconsistent and unfair consequences, because they discriminate against certain groups compared to other groups (Alper 2005: 171).

The call to distinguish genetic discrimination from other types of discrimination and subject it to special legislation has a paradoxical impact. To prohibit the "unequal treatment" of people with an "abnormal" genetic constitution reinforces the cultural belief in the exceptional status of genetic factors, something which the legal regulation was supposed to counter in the first place. The analysis and critique of genetic discrimination itself relies on the phantasm that genes provide the "blueprint" for an individual and the "secret of life."[2] In this way, the non-discrimination legislation threatens to intensify the very problem it set out to solve. This brings us up against a key dilemma (Alper 2005: 171; see also Beckwith and Alper 1998: 208). On the one hand, there are practices of genetic discrimination and people who suffer from these practices; on the other hand, the academic and legal affirmation of the exceptional role of genetic factors rejuvenates and reinforces genetic essentialism.

This dilemma by no means signifies that legal stipulations to protect people with genetic peculiarities are superfluous or even damaging. On the contrary, even if genetic information, seen scientifically, should be accorded no privileged role over non-genetic data, in social reality it is quite obviously of great importance.[3] In cultural terms, genes symbolize something fateful and immutable (Nelkin and Lindee 1995; Van Dijck 1998). They are considered "the most intimate biological property that we have" (Schmidtke 1997: 13), and are thought to decisively influence the course of an individual's life (Burnham and Phelan 2000; Little 2002). Unlike other health risks that are temporary, treatable, and can essentially be eliminated, genes are presented as the basis for a person's identity. He or she is said to "carry" or "possess" genetic risks, and these are actually considered to be an integral part of the person's physical existence (see Kavanagh and Broom 1998). If genetic factors are made responsible for a disease, then the person affected feels they are not controllable and are thus more threatening than if non-genetic reasons are cited (Senior, Marteau, and Peters 1999). Equally, the risk of genetic illness cannot be separated from the history of eugenics, the cataloging and murdering of people who were considered "genetically inferior," and from a trans-generational notion of illness and the idea of "defective" or "poor" genes (Holtzman and Rothstein 1992; Kevles 1995; Bashford and Levine 2010). As long as this cultural stereotype and historical prejudice persists, and people are disadvantaged or shown disrespect owing to their genetic peculiarities, it is imperative to legally protect them.[4]

However, such legal protection must not result in genetic data being isolated from other (predictive) medical information. Genetic discrimination is the result of an increasing extension of the concept of illness and disability and the expansion of existing practices of contempt, stigmatization, and exclusion. For this reason, prohibiting genetic discrimination must necessarily be supplemented by profound institutional reforms and comprehensive regulations that more effectively protect persons who are already ill or disabled from social exclusion and disadvantage. Otherwise we run the risk of allowing the distinction between genetic and non-genetic to lead us to ignore the more fundamental issue of how (predictive) medical information can in general be used to categorize people, to assign them characteristics and traits, and to exclude them from certain benefits (see Wolf 1995).

In the United States, therefore, the fight against genetic discrimination is increasingly seen not as an isolated problem but rather as a step on the path toward national health insurance protection which applies to everyone and will preclude differentiation on the basis of individual health risks:

> A focus on discrimination based on genetic information could conceivably draw attention away from more fundamental concerns about the health care system. One of these concerns is the limitation in availability of health insurance resulting from the use of all forms of medical testing. Thus, in the shorter term, legislation covering genetic discrimination may unintentionally serve to forestall more far-reaching reform of the health insurance system. (Alper and Beckwith 1998: 148; cf. Geetter 2002; Alper 2005)

The findings of this study show that the analysis and criticism of genetic discrimination has to be reconsidered in relation to its premises and goals. It is necessary to question two dominant narratives. The first of these considers practices of genetic discrimination to represent a radical new form of exclusion and discrimination. Its distinctive feature is that it allegedly operates at the level of the genotype and replaces visible—phenotypic—classification patterns by invisible ones. Thus, the threat supposedly arises of the creation of a new social class consisting of people who, due to their genetic characteristics, are for the most part excluded from career paths, training prospects, and opportunities for insurance, and thus socially marginalized. The problem with this critical formula lies in the way it isolates the specific factors producing inequality from one another; moreover, it is not able to examine the systematic links between already existing forms of exclusion, marginalization, and discrimination, on the one hand, and patterns of genetic explanations, on the other:

> By focusing on mechanisms of genetic discrimination and inequalities of access to genetic treatment and enhancement technologies, the use of the underclass metaphor loses much of its analytic potential. Sociologically, theories of the underclass point to cumulative dynamics of social disadvantage, geo-spatial and social forms of segregation and exclusion, and above all marginality and structural barriers to participation in the mainstream. (Kelly 2005: 147)

The dramatic invoking of the danger of a "genetic underclass" is opposed by a position that is just as popular and no less erroneous. This standpoint sees growing genetic knowledge about genetic risks and their contribution to the development of disease as promising the possibility of "genetic solidarity." This prognosis is based on the assumption that all persons "carry" "genetic errors" and thus could be affected by practices of genetic discrimination in the future.[5] The generalization and popularization of this point of view is meant to contribute to dismantling prejudicial structures, to questioning concepts of normality, and to revitalizing systems of social security and welfare (Luján and Moreno 1996). This optimistic scenario of the fundamental equality of all members of society in the face of genetic risks is not convincing for two reasons. First, it seems questionable whether rich and poor, men and women, blacks and whites are subject to practices of genetic discrimination to the same degree. It is much more probable that genetic discrimination reproduces and reinforces existing social dividing lines of race, class, and gender. The misleading view that genetic risks potentially burden all members of society equally and without distinction reflects the narrative of genetic reductionism, as it downplays the cumulative dynamics and dominance of existing patterns of social segregation and exclusion (Rouvroy 2008: 140–143). Second, what should be demonstrated is simply assumed: that new forms of identity directly and immediately follow from the knowledge of genetic risks and individual susceptibilities. This idea is debatable, because these knowledge claims are often contested and controversial. As Peter Wehling rightly stresses, "[W]hat often appears to be underestimated are the implications of the fact that such supposed biological or genetic realities are not uncontroversially given, but *defined*, *constructed* and *ascribed* by the biosciences and their specific styles of thought." (2011: 230, emphasis in original; see also Lemke 2010)

Both the dramatization and the trivialization of genetic discrimination resort to reductionist and essentialist notions of genetic regulation and systematically ignore the significance of cultural processes and social structures. The analysis and critique of genetic discrimination itself commonly rests on the fiction of a "genetic norm" which appears as an objectively given and neutral biological fact. It rarely questions this norm by examining its epistemic and social conditions of existence, and instead limits itself to improving the "treatment" of persons who deviate from the supposed genetic norm: individuals with "ill" or "defective genes" or "genetic risks."

In this manner, the analysis of genetic discrimination confirms the erroneous assumption that there are genetically normal and genetically abnormal individuals, with the additional stipulation that the latter should be treated just like the former. This kind of non-discrimination approach creates and bolsters the fiction that such a thing as a normal genotype exists; a more far-reaching approach should critically reflect on the concept of a "genetic norm" and investigate its conditions of emergence and the "truth effects" it engenders. In seeking to avoid becoming the tool of a "genetic police," the critical analysis of genetic discrimination has to take into account two aspects. It should first grasp genetic categorization and classification as an integral component of social practices producing inequality, disrespect, and stigmatization. Second, it should regard the fiction of a "genetic norm" not as an unquestioned point of departure but rather as the object of study. In the absence of such dual self-enlightenment, the critique of genetic discrimination simply leads to a further "geneticization" of the body, illness, and deviance. To conclude, the problem of genetic discrimination does not solely exist in the exceptional misuse of genetic information, but above all in its "normal" production and circulation, rather than in the exotic existence of isolated cases it materializes in the everyday routine of social practices.

Notes

NOTES TO THE INTRODUCTION

1. On Canguilhem's concept of error, see Foucault (1998).
2. In a publication on the social implications of recent genetic and reproductive technologies, Jacques Testart (2001) once mentioned the danger of a "molecular police," although he did so without further elaborating this notion.
3. *Gattaca*, Andrew Niccol's debut as a director in 1998, tells the story of Vincent who, as a "naturally conceived" being, lives the life of an outsider—of an "in-valid"—in a society whose social elite is constituted by genetically optimized people. In pursuit of his career aspiration of becoming an astronaut, he acquires the identity and body samples of a "genetically perfect" man. Living in a totalitarian surveillance society, which analyzes and tests blood, urine, skin, and hair samples on a daily basis, Vincent has to permanently fear that his inferior genetic make-up will be finally discovered. The title of the film, *Gattaca*, represents an anagram composed of the four nucleic acids of the DNA: A for adenine, C for cytosine, G for guanine, and T for thymine.

   Nancy Kress' novel *Beggars in Spain* also addresses the social gap between humans conceived the "traditional" way and their genetically improved variety. By virtue of a new method of genetic manipulation, it is possible in the future to eliminate a person's need to sleep. Millionaire Robert Camden purchases this technological innovation for his unborn daughter. Leishe, the novel's protagonist, can do entirely without sleep and is thus capable of studying and working harder than her peers. Although the "sleepless" initially meet with amusement from the majority of society, they soon occupy the most pivotal positions in business and academia due to their extraordinary capacities—a fact which soon leads to rejection, hatred, and envy on the part of the "sleepers" (Kress 1993).
4. Carol Isaacson Barash, who participated in one of the first studies on genetic discrimination herself, labelled it a "household word" (2000: 213).
5. As an illustrative example see, for instance, the circular definition provided by Gregor Wolbring: "Genetic discrimination occurs if we deal with humans or potential humans in a discriminatory fashion based on the knowledge, perception, or reality attached to the consequences of having a particular gene, gene activity, or gene product." (2005: 178)
6. This case will be discussed more thoroughly in Chapter 3.

NOTES TO CHAPTER 1

1. Etiological concepts did not always and in every case lead to a search for the cause of a disease which directs and controls a medical intervention. Thomas Schlich has shown that the idea of a "necessary cause of disease" only appeared

in the second half of the 19th century in the context of the bacteriological revolution. Whereas until then a disease had not been determined by its cause, but rather on the basis of a combination of various factors and observable characteristics, the diagnosis of a disease and the determination of its causes collapsed into each other at that time in a medicine oriented towards the natural sciences. This, however, is neither the only possible nor the only sensible foundation for medical interventions. As a result of the search for the necessary cause of disease, other factors lose significance: "The concentration on the pathogen was accompanied by the preference for the individual elimination of the necessary cause of disease and a relative lack of interest in environmental or group-related measures." (Schlich 1996: 221)

2. On the impact of genetic knowledge on medicine, see Chadarevian and Kamminga (1998), Rheinberger (1998), Ganten and Ruckpaul (2001: 6–9), Lindee (2005), and Müller-Wille and Rheinberger (2009).
3. At the end of the 1960s, the molecular biologist Robert Sinsheimer spoke of a "new horizon" within the history of humankind: "Some may smile and may feel this is but a new version of the old dream, of the perfection of man. It is that, but it is something more. The old dreams of the cultural perfection of man were always sharply constrained by his inherent, inherited imperfections and limitations . . . To foster his better traits and to curb his worse by cultural means alone has always been, while clearly not impossible, in many instances most difficult . . . We now glimpse another route—the chance to ease the internal strains and heal the internal flaws directly, to carry on and consciously perfect far beyond our present vision this remarkable product of two billion years of evolution." (Sinsheimer quoted in Keller 1992: 289–90)
4. The speech was in fact never given, since the *American Society of Human Genetics* did not convene in 1976 and took part instead in the International Congress of Human Genetics in Mexico City (cf. Childs 1977: 1).
5. In the last twenty years, there have been calls for genetic knowledge to be systematically integrated into basic medical care: "The likely increases in availability of DNA based tests and demand by patients for genetic information and advice mean that primary care practitioners will need to become genetically literate." (Emery and Hayflick 2001: 1027; cf. Caskey 1995: 155; Harris and Harris 1995)
6. Kaja Finkler points out the noticeable differences in the emphasis given to genetic causes of disease in two editions of the *Merck Manual of Diagnosis and Therapy*: "In the thirteenth edition of the manual (1977), the table of contents lacked entries referring to genetic diseases (although some genetic diseases were identified), whereas the same manual in 1997 devoted all of chapter 2 to genetics and elevated it to a separate category of 'fundamentals' absent from the 1977 edition." (2000: 14)
7. PubMed can be accessed at the following address: http://www.ncbi.nlm.nih.gov/entrez/query.fcgi
8. In the time period 2007–2012, 3,586 hits were found (accessed May 24, 2012).
9. This increase is not limited to the term "genetic risk"; similar figures are also documented for "genetic susceptibility" (29 hits between 1967 and 1976 vs. 3,181 between 1997 and 2006, and 2,820 between 2007 and 2012). However, it must be taken into account that the absolute number of entries indexed by PubMed has likewise grown during this period: from 168,310 in 1967 to 724,831in 2011 (accessed May 24, 2012).
10. Outside the medical field, some proponents of behavioral genetics even go as far as making genetic factors responsible for "social diseases" such as unemployment, alcoholism, criminality, and so on (see e.g. the literature review in

Levitt and Pieri 2009; Pieri and Levitt 2008). For an overview, see Carson and Rothstein (1999), Wasserman and Wachbroit (2001), Nuffield Council on Bioethics (2002), and Kim (2009).

11. On the problems arising from the exclusion of non-genetic factors in opting for a molecular level of explanation, see Janich (2001: 88): "If correspondingly the successful consideration of a molecular description of the human genome is reversed and a person is interpreted as being merely the realisation of a genetic plan, then obviously everything that is not genetically determined is lost. In particular, seen from an epistemological perspective, what is lost sight of is the question of how the genetically determined is to be differentiated in a methodologically clear fashion from the non-genetically determined."
12. William R. Clark holds that "ultimately *every cancer is a disorder of DNA*" (1997: 135; emphasis in original).
13. At the end of the 1990s, Rogeer Hoedemakers and Henk ten Have examined textbooks on medical genetics, official reports, and scientific articles on genetic research as well as books that aimed at presenting the newest developments of molecular genetics to a broader public. They came to the following conclusion: "In specialist textbooks, the concept of abnormality is not explained, apparently because it is taken to be self-explanatory in the given context . . . Although there is awareness of the important role of other genetic or environmental factors, the use of the concept of abnormality is often suggestive of a one-to-one correspondence between a genetic or chromosomal abnormality and clinical manifestations." (1999: 552; cf. Hoedemaekers 1998: 92–93)
14. The following considerations concentrate on two models which Smith discusses in detail in his article.
15. Cystic fibrosis is a hereditary metabolic disorder which leads to complications in breathing and digestion.
16. A similar argument is proposed by Temple (2001). However, this proposal—that is, of generally replacing the concept of genetic disease by that of genetic risk—is not convincing, because it only shifts the problem without solving it. Everything depends on the precise definition of genetic risk and the exact conceptual distinction between "disease" and (genetic) "risk."
17. On the history of the PKU test, see Chapter 3.
18. Even if all the causal factors involved in the emergence of a disease were known, the problem would still remain of the selection of those factors considered as decisive for the pathogenesis and therapy of a disease. Smith insists, in the search for the causes of a disease, on analytically distinguishing between the selection of causal factors ("causal selection") and the assumption of a "causal connection." Whereas the latter aims to systematically determine which elements within a complex set of factors can contribute to a specific outcome, the former implies a pragmatic decision: "Causal selection involves identifying which factor(s) among those known to be causally involved in a particular outcome to cite as explaining that outcome." (Smith 2001: 19)
19. Jonathan Rees therefore comes to the conclusion that genome analysis can at best supplement clinical research, but never replace it. In order to bolster his argument, he cites two examples: "Pernicious anemia is as complex a disease as one could imagine. It involves autoimmunity and complex inheritance, and it affects almost every organ system. Yet, once mechanistic insight was obtained, treatment was simple: injection of the missing vitamin B12. Successful therapy relied not on reversing the cause, because cause in this sense can be operationally defined in all sort of ways, but on finding the Achilles' heel of the condition that allowed therapeutic intervention. Similarly, syphilis, as has often been remarked, remains treatable because the microbiologists

got there before the immunologists could 'explain' the disease. The cause of disease is therefore not some objective God's eye summary of pathophysiology, but rather an operational statement of where we think the Achilles' heel of a disease might be." (2002: 699)

20. For reasons of space, I do not discuss the complex issues of whether disease can be comprehended as a purely scientific phenomenon, independent of cultural factors, and of whether the division between the subjective experience of disease and objective pathology, between "illness" and "disease," is tenable in principle. For a fundamental criticism of the biomedical approach to the problem of disease, see Morris (1998) and Fadiman (2000).
21. Compare Lloyd (1998: 555): "While a genetic classification of disease may indeed be desirable and useful, it also involves a series of judgements about the ideal forms of human life. Moving the level of diagnosis down to the molecular level does not succeed in avoiding the fundamental value-judgements involved in defining health and disease."
22. In this sense, the "process of defining something as a disease is basically a moral undertaking with moral consequences" (Freidson 1979: 283, quoted in Wiesing 1998: 96). On this issue, see also Hoedemaekers (1998: 81–88) and the contributions to the collection of articles by Brandt and Rozin (1997).
23. See Chapter 5 for an analysis of genetic reductionism as a "truth regime."
24. Antoinette Rouvroy mentions additional—and mutually reinforcing—factors which help to account for the persistence and strength of genetic reductionism: "[U]ndue extrapolations from scientific methodological reductionism, scientific strategies privileging accurately localizable genes over environmental elements that are hard to define; incentives provided by the patentability of genes and gene products; the tendency of some scientists to over-emphasise the immediate medical importance of their work to media and funding agencies; the tendency of biotechnology companies to present exaggerated claims in order to encourage their share-holders, and the reluctance to acknowledge the need for social and public remedies for individually experienced disadvantages." (2008: 35; cf. 35–54)
25. In the last decade, a broad and integral account of biological development known as "developmental systems theory" has been gaining ground in biological theory and genetic research. Proponents of this approach argue that genetic activity guides development by responding to external signals reaching specific cells at specific times. As a consequence, the dichotomy inborn vs. acquired is dispensed with and replaced by developmental patterns and forms of embodiment. Biological forms are viewed as the emergent results of interactive construction, not as manifestations of a pre-existing original plan (Oyama 2000; Oyama et al. 2001; Neumann-Held and Rehmann-Sutter 2006; see also Fausto-Sterling 2003; Stotz 2005). Paul Griffiths summarizes the approach in the following way. It encourages researchers "to investigate how a trait actually develops, what resources its reliable development depends upon, whether there are many developmental routes to this outcome, or only one, over what range of parameters is this developmental outcome stable, and how the 'environment' changes as a function of initial development differences that produce this trait." (2001: 4)
26. Compare the observation by Tom Shakespeare: "On the one foot, genetics may enable societies to avoid the birth of disabled people, and hence have an impact on the numbers of people with impairment in society, albeit not so significant an impact as some have claimed. But on the other foot, genetics has the power to turn healthy people into disabled people, by virtue of diagnosing future illness or risk of illness." (2003: 207)

NOTES TO CHAPTER 2

1. See, for instance, the proposal made by the renowned molecular biologist Linus Pauling in the late 1960s: "There should be tattooed on the forehead of every young person, a symbol showing possession of the sickle cell gene [so as to prevent] two young people carrying the same seriously defective gene in a single dose from falling in love with one another." (quoted in Markel 1992: 213)
2. See also the suggestion made by Regina Kenen: "Those diagnosed as having a late onset disorder or increased probability of developing a specific genetic disease may spend a lifetime as PPDs (Possibly Potentially Diseased) or DIWs (Diseased in Waiting)." (1996: 1550)
3. The following outline by no means claims to be exhaustive, and covers only the most important studies. For an alternative account, see Otlowski et al. (2012).
4. A short summary of the history of the Genetic Screening Study Group can be found in Alper et al. (2002: 2–4; cf. also Beckwith 2002; Alper 2005: 167–168).

   Many of its members are still active in the Council for Responsible Genetics, which was founded in the 1980s and publishes the journal *GeneWatch* (cf. Krimsky and Shorett 2005).

   For a rich account of how genetic discrimination has become a significant problem in public discourse in the United States, drawing on archive material and interviews with key actors instrumental in shaping public awareness about genetic discrimination from the 1970s onwards, see Childerhouse (2008).
5. The first results from the study were published ex ante in the journal *GeneWatch* (1991; Vol. 6, No. 4–5, p. 7, 15). For analyses of genetic discrimination conducted prior to the study by Billings et al. (1992), see Kenen and Schmidt (1978), Holtzman (1989), Duster (1990), Draper (1991), and Gostin (1991).
6. Hereditary hemochromatosis is an autosomal-recessive disease. Patients absorb an excessive amount of dietary iron from about 1 to 2 mg to about 4 to 5 mg. Excess iron cannot be excreted by the body and accumulates in tissues and organs (especially in the liver). However, the illness only manifests itself if both parents pass on the genetic mutation. Because the mutation itself rarely leads to the development of the symptoms (low penetrance), other factors probably play a role in causing the illness. On the genetic discrimination of hemochromatosis patients, see the writings of Alper et al. (1994) , Barash (2000), and Steindor (2001).
7. Symptoms may include splenic tumors, enlargement of the liver, and nervous affection. However, it may also be the case that no symptoms appear at all.
8. Those affected by Huntington's disease find themselves in a classic double-bind situation. It is often suggested to possible mutation carriers of Huntington's disease that they should not have biological children, because they could be affected by the illness (e.g., cf. Shaw 1987). Instead, adoption is generally proposed to them. In the case described above, however, this option was refused to the couple interested in adoption precisely because of the increased risk of developing the disease.
9. On the methodological shortcomings of earlier studies on genetic discrimination, see Treloar et al. (2004).
10. These diseases were chosen for the following reasons: "(1) the genetic basis of the condition is known and unambiguous; (2) discrimination directed against individuals with these conditions would most likely be due to the genetic basis of these conditions, rather than due to physical symptoms; and

(3) support groups for persons with these conditions exist so that individuals could be contacted easily. These conditions were also selected because they cover a spectrum of situations including dominant and recessive disorders, treatable and untreatable disorders, relatively common disorders for which screening programs exist, and rare disorders for which screening programs are not indicated." (Geller et al. 1996: 74)

11. See Stephenson (1992) for a personal account of genetic discrimination. Here, a mother of four had her family insurance cancelled because two of her children lived with the Fragile X syndrome, a genetic disease caused by a modified gene on the X chromosome. The disease can have a range of symptoms; it commonly results in a slowing down of mental and motor development and in behavioral disorders.
12. A survey of more than 1,500 genetic counselors and physicians conducted by Dorothy C. Wertz in the United States found that 785 patients reported having lost either their job or their insurance coverage because of genetic characteristics (Martindale 2001).
13. The conflicting results of the individual studies are likely explained by the narrower or wider definition of genetic discrimination. See the next chapter for a more detailed argument.

    Also, as some commentators have noted, Hall and Rich focused on representatives of the insurance industry and they did not interview patients themselves; this may have distorted the sample (Klitzman 2010: 69).
14. Thanks to Ine van Hoyweghen for clarifying this point to me.
15. Caution in the interpretation of the study's results is necessary, however, because a large number of those asked did not respond—possibly indicating that those applicants encountered no problems with insurers (cf. Morrison 2005: 878).
16. One case went to court in Québec in 1990: "In *Audet v. Industrielle-Alliance,* the Superior Court annulled the life-insurance contract of Mr. Tremblay, a man who had died in a car accident. Mr. Tremblay carried the genetic mutation for myotonic dystrophy of Steinert and had been informed about this. He also knew that his father and brother died of the disease. Myotonic dystrophy is a degenerative disease, which can be extremely disabling in some, while affecting others only mildly. When applying for insurance, the insurer had asked Tremblay whether he had any physical or mental anomaly. He answered no to that question. The court decided that this amounted to a false declaration, even if the disease was expressed only very mildly in him and even though his widow suggested he was 'super normal' from any point of view. As a result, the court annulled the contract *ab initio*." (Lemmens 2000: 353–354; emphasis in the original)
17. More detailed information on the overall conception of the Australian Genetic Discrimination Project is available online (http://www.gdproject.org, accessed August 20, 2012)
18. A research group led by Kristine Barlow-Stewart from the Centre for Genetics Education in Sydney has carried out a survey among 1,000 people who had undergone a predictive genetic test to determine their risk of disease. One in twelve respondents (7.3 percent) reported problems with employers or insurers. For example, one woman was found to carry the transformed gene BRCA1, which indicates an increased risk of breast and ovarian cancer. Her life insurance policy not only ruled out payment in the event of her dying of one of these two forms of cancer, but in case of death caused by any form of cancer. Barlow-Stewart stated that none of the patients filed a formal complaint or charge: "People don't know where to complain, and how to complain . . . Even though there are systems in place, they aren't using them." (Aldhous 2005)

19. Some years ago, the Council for Responsible Genetics passed a "Genetic Bill of Rights" which aims at the protection of basic human rights and the preservation of the "natural inheritance." The ten rights included in this bill encompass societal fields affected by genetic technology, from reproduction and food supply to criminal prosecution. Among the demands are, for instance, the right to a food supply that has not been genetically engineered, the right to DNA tests to defend oneself in criminal proceedings, and the right to a world in which living organisms cannot be patented. Article 8 of the catalog proclaims the right to freedom from genetic discrimination (Krimsky and Shorett 2005).
20. For an overview of the current profile of legislative and policy responses to genetic discrimination in many European and non-European countries, see Bombard et al. (2012).

    Gerards et al. (2005: 27–104) provide an older synopsis of the international status of regulation. For the regulatory approaches to genetics in the field of insurance, see the analysis by Jürgen Simon (2001) as well as the work by Bartha Maria Knoppers, Béatrice Godard, und Yann Joly (2004).
21. For an analysis of the complex relationships between governmental and non-governmental actors in Germany involved in the process of regulating genetic information, see Ireni-Saban (2010).
22. In 2005, the German National Ethics Council published a report on the reliability of predictive health information during pre-employment medical examinations. In it the council puts forward the proposal that under certain circumstances employers should be allowed to request genetic testing (National Ethics Council 2005).
23. The GEKO is an independent and interdisciplinary commission consisting of thirteen experts from different scientific backgrounds and representatives from patient groups, consumers and disabled persons. It is mentioned in Section 23 of the GenDG, and its aim is to concretize the regulations of the law and adapt it to new developments (Backhaus 2012). It is quite remarkable that through the establishment of the GEKO, the Act inscribes a continual re-assessment and re-adaptation of the GenDG.
24. A number of studies have now been produced that provide an assessment of genome analysis and genetic diagnostics with regard to the specific legal context in Germany. For a general discussion, see Hofmann (1999), Brückl (2001), Winter, Fenger, and Schreiber (2001), and Duttge, Engel, and Zoll (2011); on the issue of informational self-determination, see Stumper (1996); on the issue of the renunciation of fundamental rights in the framework of genetic diagnostics, see Eppelt (1999); on assessment in terms of constitutional law, see Meyer (2001), Tjaden (2001), and Scherrer (2012); on questions regarding insurance law, see Taupitz (2000), Simon (2001), and Ziegler and Ziegler (2011).
25. For an overview of legal protections against genetic discrimination in health insurance in the U.S., see Pollitz et al. (2007). This study presents the results of a survey among insurance companies on underwriting decisions on applications involving genetic information. The findings show that "responses did not vary when applicants were moved from states that prohibit underwriting on genetic information to states that do not" (ibid.: 365).
26. Protection against genetic discrimination guaranteed under federal law was strongly supported by public opinion (see Slaughter 2008: 727 for details). Even commercial actors and diagnostic enterprises called for a federal non-discrimination law. As the president and medical director of the U.S. genetic diagnosis provider *myDNA*, William Rice, declared, most citizens will avoid taking genetic-based tests if they do not enjoy the protection of genetic

privacy, in order to prevent discrimination at the workplace and in insurance contracts (My DNA Media 2004). In Germany, the Association of the Diagnostic Industry made a similar call in a press statement supporting legislation to secure genetic non-discrimination (Görlitzer 2004).

27. Greenwald (2010) argues that some employers might be insufficiently prepared to comply with provisions of Title II of GINA, especially in cases when employees volunteer genetic information and an adverse job action later occurs.
28. See Chapter 3 for a more detailed discussion.
29. In Germany, there is a constitutionally guaranteed principle of informational self-determination as an expression of the general personality right. This fundamental right comprises two complementary rights, the right to know and the right not to know (cf. Damm 2003). For a comprehensive discussion of "genetic privacy," see the contributions in Rothstein (1997).
30. For a comparative overview of how the different approaches are used in regulating the insurance industry in various countries, see Rothstein and Joly (2009: 139–140).

    On the complex economic, legal and social issues concerning the use of genetic information in the life insurance industry and underwriting practices and strategies, see Rothstein (2004) and van Hoyweghen (2007).
31. On this issue, see the discussion in Chapter 3.
32. Jürgen Simon has rightly pointed out that there are differences between the way the debate about genetic discrimination is conducted from the perspective of legal policy in the United States and in Europe. Whereas in the United States the term is linked to the legal instruments designed for the prevention of racial and social discrimination, in European regulation it is defined as an unjustified differential treatment of identical facts and circumstances (Simon 2001: 22; cf. also Paslack and Simon 2005: 129–130).
33. In the United Kingdom, the moratorium agreed by the government and the Association of British Insurers will expire in 2014.

NOTES TO CHAPTER 3

1. See also the comprehensive definition proposed by Therese Neuer-Miebach (2001: 54): "I term disadvantaging on the basis of a genetic predisposition that leads to disability and/or illness, or a genetically caused illness, a risk of disability or illness, genetic discrimination." (2001: 54)
2. William Nowlan and Philip R. Reilly also favor a narrow interpretation of genetic discrimination, and consider the problem's importance to have been vastly exaggerated (Nowlan 2003; Reilly 1999).
3. For a brief juxtaposition of the two poles in the discussion, see Somek (2003: 37–39).
4. For this proposed differentiation, see Hormel and Scherr (2004: 23–28); see also Gomolla and Radtke (2002).
5. See also Susan Treloar et al., who conclude that "employment and insurance have dominated the issues agenda and focus of most studies into genetic discrimination, and the defining boundaries of the phenomenon have been predetermined" (2004: 163).

    The findings of a study into genetic discrimination conducted in Australia reveal just how important it is not to focus merely on organizations. This study interviewed people who, as a result of a family history of genetic illness, were seeking genetic counseling. According to the study leader, Ms. Barlow-Steward, about 7 percent of the respondents stated they had undergone genetic testing only because of pressure from others: "The majority [of pressured cases] were from other family members. People felt pressure to assist another family member, or there was pressure to have a test because it might show something about their own health." (Robotham 2004) See also the case described by Paul Billings: "I was recently asked to consult on a case in which a woman had allegedly been repeatedly assaulted by her husband. She petitioned a court to end her marriage and grant her custody of her two young children. Her husband countered in a legal filing that she had a family history of Huntington disease, that this disorder underlay her complaints and rendered her claims for child custody inappropriate. After clinical evaluations at respected health institutions did not confirm any medical evidence of Huntington disease, the husband sought juridically ordered, forced genetic testing for the DNA expansion linked to the disorder." (2005: 560)

6. Deborah Hellman's "expressivist argument" makes a similar point: "The expressivist argument . . . focuses on the meaning expressed to everyone by the practice of genetic discrimination: does it convey the message that people with genetic conditions are less worthy of our concern and respect than others? While it may be true that an individual is pitied rather than blamed for being obese or alcoholic in that the genetic link tends to absolve individual responsibility for the condition, it may remain true that social practices which continue to treat the obese or alcoholic and others with genetic conditions less well express that the lives of these people are less valuable." (2003: 108; cf. 2008)
   For an alternative attempt at differentiation from another perspective, see Pettigrew and Taylor: "Direct discrimination occurs at points where inequality is generated, often intentionally. When decisions are based explicitly on race, discrimination is direct. Indirect discrimination is the perpetuation or magnification of the original injury. It occurs when the inequitable results of direct discrimination are used as the basis for later decisions ('past-in-present discrimination'), or decisions in linked institutions ('side-effect discrimination')." (2000: 689)
7. The term chosen here, "indirect discrimination," has to be distinguished from secondary as well as collateral discrimination. The latter refers to practices that formally espouse equal treatment, yet which have negative effects for a certain group of individuals (Röttgen 2004: 146–153), whereas the former defines the differential treatment of individuals owing to the (secondary) characteristics that impact upon them, such as level of income or education, or family status (Hormel and Scherr 2004: 26–28).
8. See the discussion in Chapter 5.
9. Gregor Wolbring rightly makes clear that such an extension of the term of genetic discrimination meets with general approval when it designates practices related to prenatal sex determination, but is less generally accepted when it refers to the prevention of characteristics regarded as disabilities or disorders. In this context, he points to a key normative tension in prenatal examinations: "On the one hand, we increase the application and usage of prebirth predictive tests for 'medical purposes' for the characteristics labeled as disabilities, diseases, or defects. And we try to prohibit the use of the same technologies for 'non-medical purposes' such as sex selection. In other words, the society believes that some characteristics deserve special protection from the misuse of the technologies (sex selection), but others don't (disability, disease, and defect deselection)." (2005: 180)
10. I employ the term "structural coupling" outside the context of systems theory and the cognitive sciences (see, for example, Luhmann 1997: 92–120); it captures very well the dynamics between operative autonomy and mutual dependence between different forms of discrimination.

11. The intimate link between genetic discrimination and discrimination against (chronically) ill and disabled persons will be discussed in chapters 4 and 5.
12. On the increasing focus on the category of "race" in genome research, see Bradby (1996), Aldhous (2002), Alper and Beckwith (2002), Nature Genetics (2004), M'charek (2005), Reardon (2005), Abu El-Haj (2007), Koenig et al. (2008), and Morning (2011).

    The danger of using genetic data in racist practices is obviously not restricted to the clinical domain. Racial discrimination can arise from the use of genetic data in criminal justice, something the inventor of the "genetic finger-print," Sir Alec Jeffreys, points to. Jeffreys has expressed concern about the number of persons included in the National DNA Database of England and Wales (which now holds more than 5 million data sets): "For a start, we are now putting not just criminals but suspects in our database, and this is clearly very highly discriminatory. If you go to certain places such as South London, you will get suspects who are predominantly black. Similarly you will get a lot of Asian suspects in Birmingham." Jeffreys concludes that the national database will be filled with a large number of blacks and persons of Asian descent who have not been found guilty of a crime, but who someone suspects will commit a crime in the future (McKie 2004; Duster 2006; Lynch et al. 2008, Hindmarsh and Prainsack 2010; M'charek, Toom, and Prainsack 2012; Heinemann, Lemke, and Prainsack 2012).
13. The term "exceptionalism" was first used in a medical context at the beginning of the 1990s in order to describe the unique type of health policy reactions to AIDS in contrast to other infectious diseases (Bayer 1991; cf. Rouvroy 2005: 196). For the term "genetic exceptionalism" and a critique, see Murray (1997).
14. See the discussion on biological complexity and genetic causation in Chapter 1.
15. I shall return to this point below. Murray terms this a "two-bucket theory of disease" that in the final instance rests on the erroneous assumption that we can make a clear medical distinction between genetic and non-genetic diseases or disease risks: "According to this model, there are two buckets—one labeled 'genetic', the other labeled 'nongenetic'—and we should be able to toss every disease and risk factor into one of the two. So, Huntington disease goes into the 'genetic' bucket and getting run over by a truck goes into the 'nongenetic' one. But many diseases and risks don't fit neatly into either bucket. Take breast cancer. Some cases of breast cancer have strong genetic roots, but others have no clear genetic connection. For that matter, not every woman with a mutated BRCA1 gene will develop breast cancer. And some apparent risk factors have little or no link to genetics. Similar complexity exists for heart disease: cholesterol is a risk factor, and one's cholesterol level can be modified by diet, exercise, and other factors; but our genes have as much or more to do with the level of cholesterol circulating in our blood as our environment or behavior. Into which bucket, then, should we toss breast cancer? Heart disease? Cholesterol level?" (Murray 1997: 67–68)
16. On the relationship between health and responsibility, see Reiser (1985), Brand and Rozin (1997), Minkler (1999), Crawshaw (2012), and Snelling (2012). See also the discussion in Chapter 5.
17. Zimmern has pointed out that the concept of genetic information possesses two mutually contradictory meanings: "First it may be regarded as information about the genetic constitution of individuals, their genes or chromosomes, and their inheritance. Second, and by contrast, genetic information may be taken to refer to any information from which we may infer knowledge about the genetic constitution of individuals." (1999: 152)

18. See, for example, Rothblatt (1997: 157): "If we have a genomic predisposition to a particular condition that interferes with our job ability, then it is wrong to deny that job until the condition manifests itself."
The American Civil Liberties Union follows a similar pattern of argumentation; in the framework of a campaign against genetic discrimination in 2002, it demanded "Congress should take immediate steps to protect genetic privacy . . . [because] it is inherently unfair to discriminate against someone based on immutable characteristics that do not limit their abilities" (quoted in Wolbring 2005: 179).
19. In fact, the distinction between asymptomatic and symptomatic individuals is more difficult to sustain than many scientists and policymakers assume. This is illustrated convincingly by the results of an empirical study based on sixty-four interviews with individuals who had, or were at risk for, Huntington's disease, breast cancer and Alpha-1 antitrypsin deficiency: "[P]atients and family members here tend not to see this differentiation as distinctly or sharply, and rather, often perceive these realms as blurring. In part, these disorders may have intermediate 'grey areas'—e.g. possible (but not definitive) and/or non-specific symptoms . . . Thus, the distinction between symptomatic vs. asymptomatic may not be wholly clear, and individuals may face discrimination due to symptoms that are not yet recognized as caused by a genetic disorder . . . Moreover, symptomatic and asymptomatic individuals are frequently members of the same families and disease communities. Fears of discrimination against symptomatic family members may heighten anxieties among their asymptomatic family members, affecting the latter's decisions." (Klitzman 2010: 80)
20. The applicant's success was limited, however. First, the court did not forbid in principle the use of genetic data as a basis for decisions on granting civil servant status, but made this into a mathematical problem of the correct calculation of risk. A (hypothetical) risk of falling ill of more than 50 percent would certainly have led to the dismissal of the lawsuit. Second, the court did not accept the applicant's claim that health data from family members should be protected from being used. The Darmstadt judges expressly allowed that in the process of conferring civil servant status, information about the genetic diseases of relatives could be asked for without their prior consent. This was justified with reference to the legitimate interest of the state in only hiring healthy applicants as civil servants. Genetic information is, according to the judgment, objectively relevant data in order to correctly determine the state of applicants' health (Ruling of the Darmstadt Administrative Court from June 24, 2004, AZ 1 E 470/04 [3]). For a critical discussion of the ruling, see Tolmein (2004).
21. Ruling of the Frankfurt am Main Administrative Court from February 2, 2004, AZ 9 G 7433/03.
22. Genetic information in the first case refers to the calculation of familial risk.
23. Zimmern rejects the issue of the technological quality of the tests as the basis for legal regulations as inappropriate; whether regulation is necessary or not, he argues, does not depend on the decision as to whether a test is genetic or non-genetic but on certain moral criteria: "The morally relevant circumstances in which regulation is warranted, whether voluntary or statutory, were in my view: (a) where there is already de facto evidence of significant discrimination or (b) where the predictive value of a test and the probability of developing the disease, or phenotypic manifestations, attributable to the genetic defect is high enough to give employers, insurers, or others in society a reason to justify discriminatory policies." (1999: 153)

24. Ramona Backhaus mentions another problematic consequence of the German GenDG: "As GenDG states it is forbidden for the insurance company to know that an insured person has undergone a genetic test—I am not talking about the results—which makes it impossible for the company to reimburse genetic tests . . . This leads to a situation that genetic tests—even if they are necessary for the individual patient—are not reimbursed by the insurance company." (2012: 123)
25. This argument is also typical of many scholarly articles that critically focus on practices of genetic discrimination. See, for example, Gostin: "Prejudice, alienation and exclusion often accompany genetically related diseases even though, by definition, the condition is neither subject to the person's control, nor the result of willful behavior." (1991: 100–111) Likewise, Rothblatt (1997: 157): "Genomic discrimination is wrong because it categorizes us on the basis of biology over which we have no control and which is irrelevant for any social, economic, or legal purpose."
26. For a similar point, see the statement of the former German Minister of Justice who holds that genes are "the most valuable good that a human being possesses" (cited by Schwägerl 2005: 4).
27. See also Beckwith and Alper: "This rationale seems to carry with it the invidious implication that we are responsible for our nongenetic diseases." (1998: 208)

    According to Alexander Somek, the phenomenon of genetic discrimination threatens the normative foundations of modern market societies. Genetic discrimination is so significant, he claims, because it reveals the limits of the liberal model and undermines the "market mentality." Liberal morality distinguishes between individual decisions on choice and fateful conditions lying outside the control of the individual. From the liberal perspective it is up to each member of society to react to blows of fate or to take measures against future unhappiness. This concept of responsibility, he explains, links social solidarity with the uncertainty about individual status of risk—a link destroyed by genetic knowledge: "Anyone can buy insurance against incidents of bad luck. Health insurance, unemployment insurance etc. offer market-based responses to the moral indifference of markets . . . Against this background, it is transparent why genetic discrimination is so awfully interesting in the context of insurance and employment contracts. Once our ill luck is known, insurance is no longer available to us. We find ourselves, in effect, bereft of the power to respond responsibly to the market's indifference toward individual responsibility." (2003: 43)

NOTES TO CHAPTER 4

1. The name of the disorder refers to the physician George Huntington, one of the first to identify and adequately describe it. The term Chorea Huntington (after the Greek word *choreia*: dance) is rarely used today, because the symptoms are not limited to movement disorders (see below).
2. The research project "Genetic Discrimination in Germany: An Investigation into Experiences of Unfair and Adverse Treatment Due to Risks of Genetic Diseases" was launched in 2011. The project will provide the first comprehensive and systematic empirical analysis of genetic discrimination in Germany. It focuses on groups of individuals with family histories, genetic risks, or genetic predispositions for four exemplary diseases: familial adenomatous polyposis (FAP), hereditary breast and ovarian cancer (BRCA1/BRCA2), iron overload disorder (hereditary hemochromatosis, HH), and cystic fibrosis (CF/mucoviscidosis). Thomas Lemke (Faculty of Social Sciences at the Goethe University

Frankfurt) and Katharina Liebsch (Department of Humanities and Social Sciences at Helmut Schmidt University Hamburg) direct the research. The project is financed by the Federal German Ministry of Education and Research (BMBF) within the priority program titled "Ethische, rechtliche und soziale Aspekte der modernen Lebenswissenschaften und der Biotechnologie" [Ethical, legal, and social aspects of modern life sciences and biotechnology]. For more information, see the website of the project: www.genetischediskriminierung.de.

3. Alleles are different variants of the same gene. Although across a population there can be several hundred different alleles in a gene, any one person has at most two different varieties in each gene locus (one each from his or her mother and father).
4. In this book the term "person at risk" serves as an abbreviation for individuals who as possible mutation carriers of HD have statistically a 50 percent risk of being affected by the disease. Even though the term has the advantage of avoiding a lengthier formulation, it should only be used cautiously. It erroneously suggests that the risk of illness is inscribed in an individual's body and corresponds to a physical state. To avoid the danger of reification the term is only used in inverted commas. With regard to the problems inherent in the loose use of the term "risk" in medical contexts, see Gigerenzer (2002). For a historical and systematic analysis of the "at-risk health status" from a sociological perspective, see Kenen (1996).
5. Recent findings indicate that the extent of the CAG expansion may have an effect on age at clinical onset (Langbehn et al. 2004; Ravina et al. 2008; see also Rosenblatt et al. 2001).
6. The ambivalences and psycho-social issues raised by the test are described in Lemke (2004a: 31–48).
7. The quotation is from the decision letter from January 3, 2000 that is in possession of the Deutsche Huntington-Hilfe (DHH).
8. The case was orally reported to the DHH board of directors.
9. The importance of data protection for those living with an HD diagnosis can hardly be overstated. To this end mail from the DHH and the *Huntington-Kurier* is dispatched to members in plain envelopes, so that it is not evident that the recipient has anything to do with HD.

   At the DHH annual conference on October 24, 2004, a young man who is a "person at risk" for HD explained to the author that the question of data protection and the fear of those affected that others might discover that they were at risk of HD was a major problem for any investigation into genetic discrimination. In his case, apart from his own family (and the self-help group), no one was aware of his risk status and DHH mail was sent to him via his parents' home address.
10. Numerous improvements were made thanks to the ideas and the dedication of various persons. Many suggestions for making the questions simpler and more precise came from Friedmar Kreuz, a member of the advisory committee and the scientific committee of the DHH, and from the chairwoman of the DHH, Christiane Lohkamp. Critical comments from Ferdinand Sutterlüty helped to revise a first version of the questionnaire.
11. One German who lived in the United States at that time filled in the questionnaire.
12. See also Andreas Kuhlmann's argument in the short report of the Bioethics Research Committee (2004).
13. The author is in possession of a copy of the letter.
14. Psychotherapeutic counseling is recommended even when the test result is negative, as knowledge that one does not "carry" the mutation radically calls into question the person's self-identity and triggers a depression in some of

those affected, for example, as a result of guilt feelings towards relatives who are ill (Tibben et al. 1997; Williams et al. 2000).

15. The "guidelines" were drawn up by a committee made up of members of the International Huntington Association and the Huntington's disease research group of the World Association for Neurology, and consist of recommendations for genetic counseling and the technical implementation of the genetic test (Medical Genetics 1994).
16. Apparently this is false information. Representatives from German insurance companies explained to the author that being in psychotherapy in the period immediately preceding an application does not necessarily lead to rejecting life insurance.
17. For further attempts at applying the concept of stigmatization to the analysis of exclusion and disrespect faced by the "potentially genetic discredited," see Miringoff (1991: 41–62), Markel (1992), Evers-Keibooms et al. (1994), Peters et al. (2005), Barclay and Markel (2007), and WonPat-Borja et al. (2012). An extensive survey of coping strategies from the perspective of social psychology is offered by Miller and Major (2000); see also Smart and Wegener (2000). The medical anthropologist Monica Konrad (2003) describes how growing genetic knowledge and improved medical technologies produce new forms of secrets and dilemmas resulting from the changed relationship to genes, bodies and the family.
18. The term used then was "Hereditary St. Vitus' Dance." On the euthanasia practice as regards HD patients, see Müller-Hill (1988) and Schmuhl (2008).
19. For the historical and theoretical limits of this perspective, see Paul (1998a) and Koch (2004).
20. The dimensions of this "genetic responsibility" are explored in the following chapter.
21. Marianne Jarka, Burkhard Brosig, and Horst-Eberhard Richter describe their experiences with 131 individuals affected directly or indirectly by HD during their five-year psychological project, which evaluated therapies for couples and families as well as group discussions with "persons at risk," patients, and relatives. The results of the study confirm fairly clear pressure on those affected to make active use of the diagnostic options available: "In our experience the desire in cases like this to find out more, or earlier, seldom occurs on the spur of the moment and on the part of the families affected. By means of what is possible with regard to genome analysis there arises subtle pressure on the families in question to make use of what is technically 'doable'. They frequently attempt to find a compromise, whereby genetic testing is only used if experienced clinical staff have been able to detect minimal symptoms or prodromes of the disease." (Brosig 1998: 35; cf. Jarka et al. 1996)
22. However, different experiences with genetic counseling and advice from medical personnel were also reported. One woman who had made use of counseling in the course of a genetic test was told by the human geneticist that he rejected abortion in the case of a positive test result, because the affected individual "in every case will be healthy for half a lifetime" (R 21).
23. For very similar reports of these "subtle, and not fully visible" forms of discrimination in employment, see Klitzman (2010: 7273).
24. Joseph Alper and Jon Beckwith use an example to illustrate this major difference between genetic and non-genetic health risks: "[M]any people who have a positive genetic test for hemochromatosis will never develop the disease, yet have been refused health insurance. Compare this use of an altered genotype as the indicator of a preexisting condition with the use of blood pressure tests. People with high blood pressure are considered to have a risk factor for heart disease, but they are not considered to have a preexisting heart condition." (1998: 146)

NOTES TO CHAPTER 5

1. In her book *A History of the Genetic Code*, the historian of science Lily Kay describes this process thus: "Discourses establish cultural efficacy through regimes of signification. These refer to the body of practices and representations a society at a particular historical period accepts and validates . . . Knowledge is therefore generated through a system of ordered procedures for the production, regulation, circulation, and operation of statements. The products of science and technology are sociotechnical; they work because they are embedded not only in material practices but also in cultural practices that stabilize and naturalize the technologies for producing knowledge and power." (2000: 18–19)
2. I refer here to the "nominalist critique" proposed by Michel Foucault (2000: 238; on this see Lemke 1997: 332–339; 2011; Dean 1998). See also the concept of a "dynamic" nominalism in Hacking (1986).
3. Cf. Hans-Jörg Rheinberger's observation: "At the beginning of the Genome Project no means were available which could have promised success for the task of sequencing the approximately three billion base pairs of the human genome. The program itself set in motion a mechanism through which these means then took shape." (1996: 293)
4. The formation of alliances and strategic coalitions here plays a decisive role by opening up new opportunities of influence, evaluation and design for the actors involved: "Molecular biology has joined forces with the human genetics counseling system, the medical-technical complex, the bio-technological industry, and forensic medicine; by doing this, all of them will change and a new medical paradigm will arise: molecular medicine." (Rheinberger 1996: 293; 292–294) On the significance of alliance building for the rise of medical genetics, see also Koch and Semerding (1994) and Gaudillière (2000).
5. According to Evelyn Fox Keller the gene discourse relies on an "epistemological culture" for which descriptions of phenomena qualify as scientific explanations, which means that they can assert themselves within the scientific system against competing interpretations (2002: 4–5).
6. In a similar manner to Kühn, Christy Simpson distinguishes three ideological positions in current health discourse, or three different forms of risk analysis and evaluation, which also determine how breast cancer and its prevention is dealt with: the belief in technological and pharmacological solutions for disease problems, the emphasis on individual health responsibility, and a social perspective, which underlines the complexity of the problem of disease and takes into account the social, economic, and political factors of the manifestation of the disease. In comparison to the two other "ideologies," however, the latter is still less accepted: "[G]iven that the technology and personal responsibility ideologies tend to have more currency in risk discourse and that, as a result, the approaches suggested by these ideologies for the prevention of breast cancer may be the ones typically expected or more readily accepted, it becomes clear that the long-term changes suggested by the social ideology will be difficult to implement." (2000: 149–150)
7. Robert Proctor's analysis of the transformation of the conception of cancer in the United States from the 1970s to the present offers an interesting illustration of a "normative etiology" and the dependence of concepts of disease on political situations. Whereas during the reform period of the 1970s medical interest concentrated on the investigation of industrially produced environmental poisons, in the Reagan era individual risk factors such as tobacco consumption and alcohol were placed at the forefront. During the 1990s, the hegemony of genetics was finally established (Proctor 1995; see also Wolf 2000). See also Joan Fujimura's thesis that the search for the causes of cancer changed in the

1980s from a "set of heterogeneous diseases marked by the common property of uncontrolled cell growth to a disease of human genes" (1996: 1).

8. Peter Conrad notes that "this shift in responsibility from society to the individual aligns with our current political climate, which increasingly blames individuals rather than social conditions for human problems. Thus genetics could become part of an ideological shift away from environmental and social analyses of problems, fostering the decline of public responsibility for human misfortune and misery." (2002: 76)
9. Pat O'Malley has diagnosed the hybrid of an "enterprising prudentialism": "The prudent subject of neo-liberalism should practice and sustain their autonomy by assembling information, material and practices together into a personalized strategy that identifies and minimizes their exposure to harm. Such risk management is frequently, and perhaps increasingly, associated with access to statistical or actuarial technologies and expert advice that render measurable the (probabilistic) calculation of future harms . . . Enterprising subjects are imagined as innovators, who 'reinvent' themselves and their environment. Here they appear as *entrepreneurs*, not as prudent *consumers* of risk . . . For the subject as entrepreneur, the future that must be governed must also remain uncertain, as a condition of a specific but vital form of liberal freedom." (2000: 465, emphasis in original)
10. Franz-Xaver Kaufmann has pointed out that the rise of the responsibility concept only began in the 1980s. "Older usage spoke of discipline, duty, liability, guilt, conscience, honesty, or morality when referring to issues for which we today use the terms responsibility or accountability. We must ask ourselves, why do such terms sound somewhat old-fashioned, or at least not appropriate when we think about typical responsibility phenomena?" (1992: 47) According to Kaufmann, the call for responsibility is loud when the recourse to "duties" alone does not suffice. In contrast to duties, where the guidelines for behavior are clear and relatively unambiguously formulated and an external, distanced implementation of behavior suffices, "responsibility" also requires an inner accordance with the behavior called for; this is to be regarded as desired and as individual will. In addition, the responsibility imperative always implies "risky decisions," in which typically latitude for judgment and alternatives for behavior are available: "Responsibility thus refers to risky tasks, i.e. tasks in which it is expected that the decision-maker will incorporate a large number of points of view, weigh various alternatives and take care to minimize possible damages. *Those tasks are regarded as responsible which require a complex estimation of the risks for their completion and therefore cannot be programmed in the form of clear rules or even directions for behavior.*" (ibid.: 45–46, emphasis in original; see also Leicht 2003)
11. Date of the searches: March 7, 2003 and May 4, 2012. The same Internet search resulted in 1,470 hits (255 in German) on February 19, 2009.
12. Date of the search: September 20, 2004. Although most of the links are now inoperative, the following discussion is not only of historical interest but will identify central elements of the discourse of genetic responsibility still valid today.
13. Oveson, L. and Yarborough, M. The Aspen Report. Ethical Issues in Occupatoinal Genetics. Available online at: http://www.ramazziniusa.org/apr01/geneticprofiles.htm (accessed September 20, 2004)
14. Fleck, L. (1997). Genetics and Ethics: A Community Dialogue. Medical Humanities Report, Vol. 18, No. 3. Available online at: http://www.bioethics.msu.edu/mhr/s97genethics.html (accessed September 20, 2004)
15. Genetic Responsibility: A Father's Part (1995). Available online at: http://media.durham.edu.on.ca/htbin/wwform/www/?TEXT=R11574881–11578061-/www/documents/037/cat/wwi770.htm (accessed September 20, 2004)

16. On the gender dimensions of genetic responsibility, see Hallowell (1999), D'Agincourt-Canning (2001), and Lemke (2004a).
17. Maxim online (2001). World O'Sex. The Sensitivity Card. Available online at: http://www.maximonline.com/world_o_sex/article_3952.html (accessed September 20, 2004)
18. Science Generation. Genetics and Polemics. Available online at: http://en.science-generation.com/biomgene.xml (accessed September 20, 2004)
19. For a discussion of "genetic irresponsibility" from a bioethical perspective, see Andre, Fleck, and Tomlinson (2000).
20. In February 1975, 150 scientists met in Asilomar to discuss the possible dangers of genetic technologies, a new question at that time. In their concluding paper, a large majority called for the establishment of mandatory guidelines for the carrying out of genetic experiments in order to guarantee a responsible use of the new technologies (Gottweis 1998: 87–105).
21. See, for example, Dan W. Brock's observation that the results of the Human Genome Project are "likely to affect deeply . . . our conception of ourselves as responsible agents and, more specifically, as morally and legally responsible for our actions, for the lives we live, and for the kinds of people that we become" (1999: 23; see also Somek 2003).
22. Of course, the idea of a "genetic responsibility" has existed much longer. The concept of responsible parenthood was articulated in the 19th century by the liberal John Stuart Mill: "The fact itself, of causing the existence of human life, is one of the most responsible actions in the range of human life. To understand this responsibility—to bestow a life which may be either a curse or a blessing—unless the being on whom it is to be bestowed will have at least the ordinary chances of a desirable existence, is a crime against that being." (Mill 1963: 133)
23. Lebel (1977: 67A) approaches the problem of genetic responsibility in a similar manner, distinguishing between two complexes of questions: "(1) to whom are parents answerable in their decision-making process?, and (2) what considerations constitute an adequate ('responsible') decision-making process?"
24. In the following, I limit myself to those "new" facets of genetic responsibility and exclude, for the most part, the analysis of "reproduction responsibility." For analyses of the responsibility discourse in prenatal diagnostics, see Marteau and Drake (1995), Beck–Gernsheim (1996), and Ettore (2002).
25. One could say that here there is no responsibility for the problem, but there is a responsibility for its solution. In his analysis of the empowerment concept, Ulrich Bröckling has pointed out that the invocation of self-responsibility can in fact be (initially) part of eliminating "victim blaming": "The motto for this is provided by the civil rights activist Jesse Jackson: 'You are not responsible for being down, but you are responsible for getting up'." (2003: 333)
26. Anne Kerr and Tom Shakespeare likewise distinguish two forms of responsibility which can arise from the diagnosis of genetic risks: "The first is to avoid behaviours likely to exacerbate that risk. This starts with consulting and following the advice of medical experts . . . Second, individuals bear responsibility for informing their genetic kin about their risk." (2002: 153–154)
27. As Robert Edwards, the pioneer of in vitro fertilization (IVF), once put it: "In the future, it will be a sin to have a disabled child." (quoted by Shakespeare 2003: 205)
28. For an analysis of the shift in emphasis and the expansion of the concept of genetic responsibility in the medical advice literature of the last twenty-five years, see Lemke (2006).
29. It is important to note that the problem of data protection, confidentiality and privacy goes far beyond the clinical context. It is a central issue in

biomedical research, the operation of biobanks and the uptake of direct-to-consumer genetic testing, all of which pose new challenges in safeguarding genetic privacy and non-discrimination (Greely 2007; Prainsack 2008; Heeney et al. 2010).

30. For discussions of the legal obligation of the physician to pass on genetic diagnosis results to relatives in the German legal system, see Cramer (1991: 201–279), Stumper (1996: 218–234), Eppelt (1999: 236–285), Hofmann (1999: 139–152), and Meyer (2001: 102–142).
31. See also the position paper of the Deutsche Gesellschaft für Humangenetik (German Society of Human Genetics), which states that there is "a moral obligation . . . to share genetic knowledge" in families if there are preventive, therapeutic or pre-natal diagnostic options available (2007: 7–8).
32. The professional obligation that physicians could be liable to persons with whom they have never established a patient–physician relationship was stated in an earlier lawsuit (*Tarasoff et al. v. The Regents of the University of California et al.* 1976 [17 Cal.3d 425]). On prior lawsuits in which the issue of information on genetic risks was raised, see Andrews (1997: 266–273); see also Deftos (1998).
33. See *Pate v. Threlkel* (1995: 282): "The patient ordinarily can be expected to pass on the warning. To require the physician to seek out and warn various members of the patient's family would often be difficult or impractical and would place too heavy a burden upon the physician. Thus, we emphasize that in any circumstances in which the physician has a duty to warn of a genetically transferable disease, that duty will be satisfied by warning the patient."
34. For a more comprehensive overview of how the intra-familial disclosure and familial management of genetic information is dealt with in court decisions in the United States and other countries, see Rouvroy (2008: 71–79).
35. The following distinction draws on the argument in the ASHG statement (1998).
36. Among other reasons, the judge in the first court had dismissed the case because of this special quality of genetic risks: "The harm is already present within the non-patient child as opposed to being introduced, by a patient who was not warned to stay away." (677 A.2d 1188, N.J. Super. [1996]: 624, cited in Andrews 1997: 268)
37. Informational and reproductive responsibility intersect at this point—with extremely serious consequences, as Lori Andrews remarks: "[C]onsideration should also be given to the eugenic symbolism of breaching confidentiality to warn relatives. The National Research Council's Committee for the Study of Inborn Errors of Metabolism cautioned that allowing a breach of confidentiality to warn relatives without consent implies 'a medical-social judgment that genetic "normality" is a prime childbearing goal for society to follow'." (1997: 272)
38. Brian R. Gin assumes a structural similarity as regards disease and disability status between people with a genetic risk for HD and individuals with a positive AIDS test result: "*Inter alia*, Huntington's disease and HIV-positive status are analogous in that, at the time of diagnosis, victims of both diseases may have no symptoms and may remain healthy for a number of years; but even though the exact time of onset of both diseases is unascertainable, death of both victims within a given range of years is highly likely. Further, both Huntington's disease and HIV are transmitted to offspring at a relatively high rate." (1997: 1406)
    Christine Scholz (1995: 50) mentions that the belief that sickle cell anaemia was infectious was widespread in the United States until the 1980s.
39. The following argument is based primarily on Petrila (2001: 411–418).

40. Regina Kenen points out that the "geneticization" of familial relations could also give rise to new obligations to inform even before any decision to start a family: "[T]he norm for the sharing of private genetic information might lead to 'genetic cheating', a new form of deviance in conjugal relationships. The Catholic church now grants annulments when prior to the marriage one prospective partner lies or withholds crucial information from the other that might have influenced the decision to marry. In the future, would one basis for annulments (or divorces, in the case of other religions) include hiding genetic information or refusing to be tested if one partner knew that she might be at more than average risk?" (1994: 58; cf. Finkler 2000)
41. As a genetic counselor put it, "Each of us has the ethical responsibility to communicate vital information to our relatives . . . Hiding the truth behind a cover of concern for the feelings of individual family members is unacceptable when the wellness and the very life of another [are] at stake. We have a moral imperative to tell." (Milunsky 2001: 7; for a critique of this argument see Lemke 2006)
42. For a similar scenario, see the Internet publication "Your Genes, Your Choices," by the American Association for the Advancement of Science (2003), which emphasizes the importance of "genetic literacy." For more complex accounts of the question of whether hypothetical genetic testing intentions are good proxies for changes in health behavior, see Sanderson et al. (2010) and McBride et al. (2010).
43. For an elaboration of this argument, see Lemke (2004b).
44. See also Irrgang: "The genetic interpretation of many causes of disease in predictive medicine has led to a series of diagnoses of diseases before symptoms have ever appeared. These are not diseases in the strict sense, but actually dispositions towards a disease. Patients in predictive medicine are thus not in fact ill persons, but people who want to find out about genetic information which could be medically relevant. The potentially ill thus receive a new responsibility for their getting sick in the future." (2001: 652)
45. For a documentation of the decision, see http://www.oyez.org/cases/2000–2009/2001/2001_00_1406 (accessed on August 30, 2012).
46. The EEOC is a parastatal commission promoting equal opportunities at the workplace and the prevention of discrimination against employees.
47. Susceptibility means the individual's "receptivity" to a disease or to the effects of harmful substances.
48. Olivier De Schutter (2004: 18–22) demonstrates that the directives of EU law make a similar decision in the member states of the European Union probable. For discussions of the German legal situation as regards genetic tests in employment, see Roos (1999), Meyer (2001), Sonnenfeld (2002), Zinke (2003), and Tzortzis (2004). On "Admissibility of Genome Analyses on Employees in German and US Law," see the—somewhat older—legal comparison by Pletke (1997).
49. See the statement by Nora Kizer Bell: "[I]f a person voluntarily chooses to perform an action that is harmful to self, this is taken to be sign of an inability to act responsibly. The act is regarded as not voluntary in the crucial sense and taken to signal that the individual is not exercising autonomous choice." (1996: 779)

NOTES TO CHAPTER 6

1. Such cases concern the birth of a disabled child which, according to the mother or the parents, would have been aborted if medical advice before or during the pregnancy had not been erroneous or insufficient. The error

can lie in genetic consultation or a prenatal diagnosis that was mistakenly omitted or incorrectly conducted. In these cases, the parents or the mother successfully sued the physician for financial support for the child and for compensation for the period of pregnancy and birth. Although the claim for damages was based on insufficient or erroneous medical care, the basis for the claim is still the "genetic damage" of the child (Neuer-Miebach 2001: 57; Degener 1998; Wolbring 2001: 88–94; Steinbock 2009; Stein 2010).

2. Ingrid Schneider points to a similar problem in the context of the debate on patenting human genes: "[B]ecause critics and the persons affected assert that the patenting robs them of their innermost being or 'identity', they reproduce and confirm a genetic determinism and essentialism that erroneously reduces human existence to DNA structures." (1999: 15)
3. This argument is presented in detail by Deborah Hellman: "This article will develop the argument that because the social meaning of treating people differently on the basis of their genetic make-up is different from the social meaning of discrimination on the basis of health or illness, special legislation is warranted to prohibit genetic discrimination." (2003: 79; see also Murray 1997: 71) See also the following argument by Keays (2000: 83): "[S]imply because genetic test information does not possess some special quality does not lead to the conclusion that insurance companies should have a free rein in discriminating on the basis of genetic test results. Indeed it is largely irrelevant. What is important are the social implications of discriminating on the basis of genetic tests."
4. This point marks the main difference between the analysis being put forward here and the position taken by Ulrich Stockter, who also emphasizes that a prohibition of genetic discrimination is counterproductive in terms of legal policy as it fosters the genetic determinism it seeks to fight. At the same time, Stockter claims that "statements based on genetic features are not of special importance" (2004: 41), which is why, in his view, the unrestricted use of predictive data in the insurance industry is acceptable and legitimate.
5. "[E]ach one of us [is] the carrier of one or another serious hereditary disease . . . Moreover, every person has hereditary changes that predispose one to one or more diseases. For this reason, the distinction between the hereditary 'healthy' and 'ill' is not sensible." (Bartram et al. 2000: 58; cf. Henn 2001; Shakespeare 2003: 206)

Bibliography

Abdulrahman, G.O. (2011). Therapeutic advances in the management of Huntington's disease. *Yale Journal of Biology and Medicine* 84: 311–319.

Adams, M. (1990). *The wellborn science: Eugenics in Germany, France, Brazil, and Russia*. New York: Oxford University Press.

Agar, N. (1998). Liberal eugenics. *Public Affairs Quarterly* 12(2): 137–156.

Agar, N. (2004). *"Liberal eugenics": In defence of human enhancement*. Oxford: Blackwell.

Aldhous, P. (2002). Geneticist fears "race-neutral" studies will fail ethnic groups. *Nature* 418: 355–356.

Aldhous, P. (2005). Victims of genetic discrimination speak up. *New Scientist* November 2, 2005. Available online at: http://www.eurekalert.org/pub_releases/2005-11/ns-vog110205.php (accessed August 24, 2012).

Allen, A.L. (1997). Genetic privacy: Emerging concepts and values, pp. 31–59 in M.A. Rothstein (ed) *Genetic secrets: Protecting privacy and confidentiality in the genetic era*. New Haven, CT: Yale University Press.

Alper, J.S. (2005). Beyond genetic anti-discrimination, pp. 167–172 in S. Krimsky and P. Shorett (eds) *Rights and liberties in the biotech age: Why we need a genetic bill of rights*. Lanham, MD: Rowman and Littlefield.

Alper, J.S., C. Ard, A. Ash, P. Conrad, L.N. Geller, and J. Beckwith (eds). (2002). *The double-edged helix: Social implications of genetics in a diverse society*. Baltimore: John Hopkins University Press.

Alper, J.S. and J. Beckwith. (1998). Distinguishing genetic from nongenetic medical tests: Some implications for antidiscrimination legislation. *Science and Engineering Ethics* 4: 141–150.

Alper, J.S. and J. Beckwith. (2002). Genetics, race, and ethnicity, pp. 175–196 in J.S. Alper, C. Ard, A. Ash, P. Conrad, L.N. Geller, and J. Beckwith (eds) *The double-edged helix: Social implications of genetics in a diverse society*. Baltimore: John Hopkins University Press.

Alper, J.S., L.N. Geller, C.I. Barash, P.R. Billings, V. Laden, and M.R. Natowicz. (1994). Genetic discrimination and screening for hemochromatosis. *Journal of Public Health Policy* 15: 345–358.

American Society of Human Genetics Social Issues Subcommittee on Familial Disclosure. (1998). ASHG statement: Professional disclosure of familial genetic information. *American Journal of Human Genetics* 62: 474–483.

Andre, J., M.L. Fleck, and T. Tomlinson. (2000). On being genetically "irresponsible". *Kennedy Institute of Ethics Journal* 10(2): 129–146.

Andrews, L.B. (1997). Gen-etiquette: Genetic information, family relationships, and adoption, pp. 255–280 in M.A. Rothstein (ed) *Genetic secrets: Protecting privacy and confidentiality in the genetic era*. New Haven, CT: Yale University Press.

Andrews, L.B. (2001). *Future perfect: Confronting decisions about genetics*. New York: Columbia University Press.

Appelbaum, P.S. (2010). Genetic discrimination in mental disorders: The impact of the genetic information nondiscrimination act. *Law and Psychiatry* 61(4): 338–340.

Armstrong, D., S. Michie, and T. Marteau. (1998). Revealed identity: a study of the process of genetic counselling. *Social Science & Medicine* 47: 1653–1658.

Backhaus, R. (2011). No magic bullet: The impact of the Genetic Diagnostic Act on the private insurance market in Germany, pp. 153–166 in K. Horstman, E. Dow, and B. Penders (eds) *Governance of health care innovation: Excursions into politics, science and citizenship*. Raleigh, NC: Lulu Academic.

Balding, D. and S. Richardson. (2000). Special issue: Modern statistical methods for disease gene mapping. *International Statistical Review* 68: 1–3.

Barad, K. (2007). *Meeting the universe halfway: Quantum physics and the entanglement of matter and meaning*. Durham, NC: Duke University Press.

Barash, C.I. (2000). Genetic discrimination and screening for hemochromatosis: Then and now. *Genetic Testing* 4: 213–218.

Barclay, L.A. and K.S. Markel. (2007). Discrimination and stigmatization in work organizations: A multiple level framework for research on genetic testing. *Human Relations* 60(6): 953–980.

Barlow-Stewart, K. and D. Keays. (2001). Genetic discrimination in Australia. *Journal of Law and Medicine* 8: 250–262.

Barlow-Stewart, K., S. Taylor, S. Treloar, M. Stranger, and M. Otlowski. (2009). Verification of consumers' experiences and perceptions of genetic discrimination and its impact on utilization of genetic testing. *Genetics in Medicine* 11(3): 193–201.

Bartram, C.R., J.P. Beckmann, F. Breyer, G. Fey, C. Fonatsch, and D. Uhl (eds) (2000). *Humangenetische Diagnostik. Wissenschaftliche Grundlagen und gesellschaftliche Konsequenzen*. Berlin: Springer.

Bashford, A. and Philippa L. (eds) (2010). *The Oxford handbook of the history of eugenics*. New York: Oxford University Press.

Bauer, S. (2008). Societal and ethical issues in human biomonitoring? A view from science studies. *Environmental Health* 7(1): 10.

Bauer, S. (2009). From society to molecule and back: The contested scale of public health science, pp. 113–134 in S. Bauer and A. Wahlberg (eds) *Contested categories: Life sciences in society*. Farnham: Ashgate.

Bause, M. (2000). Guter Rat ist teuer—humangenetische Beratung unter den Bedingungen der Marktindividualisierung, pp. 96–106 in J. Schmidtke (ed) *Guter Rat ist teuer. Was kostet die Humangenetik, was nutzt sie?* München: Urban & Fischer.

Bayer, R. (1991). Public health policy and the AIDS epidemic: An end to HIV exceptionalism? *New England Journal of Medicine* 21(324): 1500–1504.

Beck-Gernsheim, E. (1996). Die soziale Konstruktion des Risikos—das Beispiel Pränataldiagnostik. *Soziale Welt* 47: 284–296.

Beckwith, J. (2002). *Making genes, making waves*. Cambridge, MA: Harvard University Press.

Beckwith, J. and J.S. Alper. (1998). Reconsidering genetic antidiscrimination legislation. *Journal of Law, Medicine and Ethics* 26: 205–210.

Beeson, D. and T. Doksum. (2001). Family values and resistance to genetic testing, pp. 153–178 in B. Hoffmaster (ed) *Bioethics in social context*. Philadelphia: Temple University Press.

Bell, J. (1998). The new genetics: The new genetics in clinical practice. *British Medical Journal* 316(7131): 618–620.

Bell, N.K. (1996). Responsibilities and rights in the promotion of health: Differing positions of the individual and the state. *Social Science & Medicine* 43(5): 775–782.

Bergsma, D. (ed) (1974). *Ethical, social and legal dimensions of screening for human genetic disease.* New York: Stratton Intercontinental Medical Book Corporation.

Billings, P.R. (1991). Genetic discrimination: An ongoing survey. *GeneWatch* 6(4–5): 7–15.

Billings, P.R. (2005). Genetic nondiscrimination. *Nature Genetics* 37(6): 559–560.

Billings, P.R., M.A. Kohn, M. de Cuevas, J. Beckwith, J.S. Alper, and M.R. Natowicz. (1992). Discrimination as a consequence of genetic testing. *American Journal of Human Genetics* 50: 476–482.

Bioethics Research Committee. (2004). *Expert panel discussion organised by the Forschungsstelle Bioethik and Institut für Ethik, Theorie und Geschichte der Medizin on June 29, 2004 in Münster on 'genetische Diskriminierung'.* Available online at: http://www.uni-muenster.de/imperia/md/content/bioethik/archiv/kurzbericht_gen_diskr.pdf (accessed June 27, 2012).

Bodie, M. (2005). Eddy Curry and genetic discrimination. *PrawfsBlawg*, July 2005. Available online at: http://prawfsblawg.blogs.com/prawfsblawg/2005/10/the_notion_of_g.html (accessed August 24, 2012).

Bombard, Y., E. Penziner, J. Decolongon, M.L.N. Klimek, S. Creighton, O. Suchowersky, M. Guttman, J.S. Paulsen, J.L. Bottorff, and M.R. Hayden. (2007). Managing genetic discrimination: Strategies used by individuals found to have the Huntington disease mutation. *Clinical Genetics* 71(3): 220–231.

Bombard, Y., E. Penziner, O. Suchowersky, M. Guttman, J.S. Paulsen, J.L. Bottorff, and M.R. Hayden. (2008). Engagement with genetic discrimination: Concerns and experiences in the context of Huntington disease. *European Journal of Human Genetics* 16(3): 279–289.

Bombard, Y., J. Palin, J.M. Friedman, G. Veenstra, S. Creighton, J.L. Bottorff, M.R. Hayden, and Canadian Respond-HD Collaborative Research Group. (2012). Beyond the patient: The broader impact of genetic discrimination among individuals at risk for Huntington disease. *American Journal of Medical Genetics Part B. Neuropyschiatric Genetics* 159B(2): 217–226.

Bombard, Y., J. Palin, J.M. Friedman, G. Veenstra, S. Creighton, J.S. Paulsen, J.L. Bottorff, M.R. Hayden, and Canadian Respond-HD Collaborative Research Group. (2010). Factors associated with experiences of genetic discrimination among individuals at risk for Huntington disease. *American Journal of Medical Genetics Part B. Neuropsychiatric Genetics* 156(1): 19–27.

Bradby, H. (1996). Genetics and racism, pp. 295–316 in T. Marteau and M. Richards (eds) *The troubled helix: Social and psychological implications of the new human genetics.* Cambridge: Cambridge University Press.

Brand, A., H. Brand, and T. Schulte in den Baumen. (2008). The impact of genetics and genomics on public health. *European Journal of Human Genetics* 16(1): 5–13.

Brandt, A.M. and P. Rozin. (1997). *Morality and health.* New York: Routledge.

British Medical Association. (1998). *Human genetics: Choice and responsibility.* Oxford: Oxford University Press.

Brock, D.W. (1999). The Human Genome Project and human identity, pp. 18–33 in R.F. Weir, S.C. Lawrence, and E. Fales (eds) *Genes and human self-knowledge: Historical and philosophical reflections on modern genetics.* Iowa City: University of Iowa Press.

Bröckling, U. (2003). You are not responsible for being down, but you are responsible for getting up. Über Empowerment. *Leviathan* 31: 323–344.

Bröckling, U., S. Krasmann, and T. Lemke. (eds) (2011) *Governmentality: Current issues and future challenges.* New York: Routledge.

Brosig, B. (1998). Compassion: Meine Erfahrung mit Chorea Huntington. *Psychosozial* 21: 27–35.

Browaeys, D.B. and J.-C. Kaplan. (2000). Die Versuchung der genetischen Apartheid. *Le Monde Diplomatique* May 12, 2000: 1, 12–13.
Brückl, D. (2001). *Rechtsfragen zur Verwendung von genetischen Informationen über den Menschen. Ein Beitrag zur rechtlichen Steuerung der Verwendung von Wissen.* Berlin: Weißensee Verlag.
Buchanan, A., D. Brock, N. Daniels, and D. Wikler. (2000). *From chance to choice: Ethics and justice.* Cambridge: Cambridge University Press.
Bulmahn, E. (2003). Vorwort, pp. 6–7 in Bundesministerium für Bildung und Forschung (ed) *Krankheitsbekämpfung durch Genomforschung. Das Nationale Genomforschungsnetz.* Bonn: BMBF.
Bundesärztekammer. (2003). Richtlinien zur prädiktiven genetischen Diagnostik. *Deutsches Ärzteblatt* 100(19): A 1297–1305.
Bundesministerium für Gesundheit und Soziale Sicherung. (2004). *Begründung zum Diskussionsentwurf "Gesetz über genetische Untersuchungen bei Menschen."* Berlin: BMGS.
Bunton, R. and A. Petersen. (2005). *Genetic governance: Health, risk and ethics in the biotech era.* London: Routledge.
Burgermeister, J. (2003). Teacher was refused job because relatives have Huntington's disease. *British Medical Journal* 327(7419): 827a.
Burnham, T. and J. Phelan. (2000). *Mean genes. From sex to money to food: Taming our primal instincts.* Cambridge, MA: Perseus.
Callon, M. and V. Rabeharisoa. (2004). Gino's lesson on humanity: Genetics, mutual entanglements and the sociologist's rule. *Economy and Society* 33: 1–27.
Canguilhem, G. (1978). *On the normal and the pathological.* Dortrecht: Reidel.
Carson, R.A. and M.A. Rothstein. (1999). *Behavioral genetics: The clash of culture and biology.* Baltimore: John Hopkins University Press.
Caskey, C.T. (1995). Medizin auf der Grundlage der DNA—Prävention und Therapie, pp. 123–155 in D.J. Kevles and L. Hood (eds) *Der Supercode. Die genetische Karte des Menschen.* Frankfurt am Main: Insel.
Castel, R. (1991). From dangerous to risk, pp. 281–297 in G. Burchell, C. Gordon, and P. Miller (eds) *The Foucault effect: Studies in governmentality.* Hemel Hempstead, UK: Harvester Wheatsheaf.
Chadarevian de, S. and H. Kamminga (eds). (1998). *Molecularizing biology and medicine: New practices and alliances, 1910s–1970s.* Amsterdam: Overseas Publishers Association.
Childerhouse, J.E. (2008). *Genetic discrimination: Genealogy of an American problem* (PhD diss., McGill University, Montreal).
Childs, B. (1977). Persistent echoes of the nature–nurture argument. *American Journal of Human Genetics* 29: 1–13.
Childs, B. (1999). *Genetic medicine: A logic of disease.* Baltimore: John Hopkins University Press.
Christiaans, I., K.T.M Kok, I.M. van Langen, E. Birnie, G.J. Bonsel, A.A. Wilde, and E.M. Smets. (2010). Obtaining insurance after DNA diagnostics: A survey among hypertrophic cardiomyppathy mutation carriers. *European Journal of Human Genetics* 18(2): 251–253.
Clark, W.R. (1997). *The new healers: The promise and problems of molecular medicine in the twenty-first century.* New York: Oxford University Press.
Clarke, A. and A. Lucassen (2007). Should families own genetic information? No. *British Medical Journal* 335: 23.
Clayton, E.W. (2003). Ethical, legal, and social implications of genomic medicine. *New England Journal of Medicine* 349: 562–569.
Coalition for Genetic Fairness. (2008). *What does GINA Mean? A guide to the Genetic Information Nondiscrimination Act.* Available online at: http://www.geneticfairness.org/GINAPublication111008.pdf (accessed August 24, 2012).

Collins, F.S. (1999). Shattuck lecture—medical and societal consequences of the Human Genome Project. *New England Journal of Medicine* 341: 28–37.

Conrad, P. (1999). A mirage of genes. *Sociology of Health & Illness* 21(2): 228–241.

Conrad, P. (2002). Genetics and behavior in the news: Dilemmas of a rising paradigm, pp. 58–79 in J.S. Alper, C. Ard, A. Ash, P. Conrad, L.N. Geller, and J. Beckwith (eds) *The double-edged helix: Social implications of genetics in a diverse society*. Baltimore: John Hopkins University Press.

Cramer, S. (1991). *Genom- und Genanalyse. Rechtliche Implikationen einer "prädiktiven Medizin."* Frankfurt am Main: Peter Lang.

Crawford, R. (1977). You are dangerous to your health: Ideology and politics of victim blaming. *International Journal of Health Services* 7(4): 663–680.

Crawshaw, P. (2012). Governing at a distance: Social marketing and the (bio)politics of responsibility. *Social Science and Medicine* 75: 200–207.

Cruikshank, B. (1999). *The will to empower: Democratic citizens and other subjects*. Ithaca, NY: Cornell University Press.

D'Agincourt-Canning, L. (2001). Experiences of genetic risk: Disclosure and the gendering of responsibility. *Bioethics* 15(3): 231–247.

Damm, R. (1999). Prädiktive Medizin und Patientenautonomie. Informationelle Patientenrechte in der Gendiagnostik. *Medizinrecht* 10: 437–448.

Damm, R. (2003). Prädiktive genetische Tests: Gesellschaftliche Folgen und rechtlicher Schutz der Persönlichkeit, pp. 203–228 in L. Honnefelder, D. Mieth, P. Propping, L. Siep, and C. Wiesemann (eds) *Das genetische Wissen und die Zukunft des Menschen*. Berlin: Gruyter.

Daniels, N. (2003). Chevron v Echazabal: Protection, opportunity, and paternalism. *American Journal of Public Health* 93: 545–548.

De Schutter, O. (2004). *Pre-employment inquiries and medical examinations as barriers to the employment of persons with disabilities: Reconciling the principle of equal treatment and health and security regulations under European Law*. Available online at: http://edz.bib.uni-mannheim.de/daten/edz-ath/gdem/04/disabheasafe.pdf (accessed August 25, 2012).

Dean, M. (1998). Questions of method, pp. 182–199 in I. Velody and R. Williams (eds) *The politics of constructionism*. London: Sage.

Dearing, A. (2002). Genetic discrimination. *Risk Management* 49(7): 8–9.

Deftos, L.J. (1998). The evolving duty to disclose the presence of genetic disease to relatives. *Academic Medicine* 73: 962–968.

Degener, T. (1998). Ein behindertes Kind als Schaden? *Gen-Ethischer Informationsdienst* 129: 26–31.

Der Spiegel. (2004). Gentest für Job-Suchende. *Der Spiegel* 42: 19.

Deutsche Huntington-Hilfe. (2001). Wir über uns, pp. 57–60 in Deutsche Huntington-Hilfe (ed) *Huntington-Krankheit. Eine Informationsschrift*. Duisburg: Eigenverlag.

Deutscher Bundestag. (2002). *Schlussbericht der Enquete-Kommission "Recht und Ethik in der modernen Medizin."* Opladen: Leske und Budrich.

Dickens, P. (2001). Linking the social and natural sciences: Is capital modifying human biology in its own image? *Sociology* 35: 93–110.

Dickinson, B.D., M. Genel, C.B. Robinowitz, P.L. Turner, and G.L. Woods. (2002). Gender verification of female Olympic athletes: Medicine and science. *Sports and Exercise* 34: 1539–1542.

Diver, C.S. and J.M. Cohen. (2001). Genophobia: What is wrong with genetic discrimination? *University of Pennsylvania Law Review* 149: 1439–1482.

Dose, M. (1997). Klinische Diagnostik und Therapie bei Anlageträgern der Huntington-Krankheit. *Medizinische Genetik* 9: 570–579.

Douglas, M. (1990). Risk as a forensic resource. *Daedalus* 119(4): 1–16.

Dowbiggin, I.R. (2008). *The sterilization movement and global fertility in the 20th century*. Oxford: Oxford University Press.

Draper, E. (1991). *Risky business: Genetic testing and exclusionary practices in the hazardous workplace*. Cambridge: Cambridge University Press.

Duden, B. (2002). *Die Gene im Kopf, der Fötus im Bauch*. Hannover: Offizin

Duster, T. (1990). *Backdoor to eugenics*. New York: Routledge.

Duster, T. (2006). The molecular reinscription of race: Unanticipated issues in biotechnology and forensic science. *Patterns of Prejudice* 40(4–5): 427–441.

Duttge, G., W. Engel, and B. Zoll. (eds). (2011). *Göttinger Schriften zum Medizinrecht Bd.11. "Das Gendiagnostikgesetz im Spannungsfeld von Humangenetik und Recht."* Göttingen: Universitätsverlag Göttingen.

Eberbach, W. (2010). Das neue Gendiagnostikgesetz. Ein Überblick aus juristischer Sicht. *Medizinrecht* 28(3): 155–163.

Emery, J. and S. Hayflick. (2001). The challenge of integrating genetic medicine into primary care. *British Medical Journal* 322: 1027–1030.

The ENCODE Project Consortium. (2007). Identification and analysis of functional elements in 1% of the human genome by the ENCODE pilot project. *Nature* 447: 799–816.

Enquete-Kommission Recht und Ethik der modernen Medizin. (2000). *Öffentliche Anhörung von Sachverständigen zum Thema "Folgen der genetischen Diagnostik" (Protokoll)*. Available online at: http://www.roespel.de/medi_oe2.pdf (accessed December 13, 2012).

Environmental Protection Agency. (2004). *Potential implications of genomics for regulatory and risk assessment applications at EPA* (Report prepared for the U.S. Environmental Protection Agency by members of the Genomics Task Force Workgroup, a group of EPA's Science Policy Council). Available online at: http://www.epa.gov/osa/pdfs/EPA-Genomics-White-Paper.pdf (accessed August 25, 2012).

Eppelt, M.D. (1999). *Grundrechtsverzicht und Humangenetik. Der Verzicht auf Grundrechte, insbesondere im Rahmen der Einwilligung in die Anwendung neuerer, humangenetischer Diagnose- und Therapieformen*. Herdecke: GCA-Verlag.

Ericson, R., D. Barry, and A. Doyle. (2000). The moral hazards of neo-liberalism: Lessons from the private insurance industry. *Economy & Society* 29(4): 532–558.

Ettorre, E. (2002). A critical look at the new genetics: Conceptualizing the links between reproduction, gender and bodies. *Critical Public Health* 12(3): 237–250.

European Association for the Study of the Liver. (2010). EASL clinical practice guidelines for HFE hemochromatosis. *Journal of Hepatology* 52: 3–22.

European Commission (ed) (2004). *Ethical, legal and social aspects of genetic testing: Research, development and clinical applications*. Brussels: EC.

Evers-Kiebooms, G., L. Denayer, M. Welkenhuysen, J.-J. Cassiman, and H. Van den Berghe. (1994). A stigmatizing effect of the carrier status for cystic fibrosis. *Clinical Genetics* 46: 336–343.

Fadiman, A. (2000). *Der Geist packt dich und du stürzt zu Boden*. Berlin: Berlin Verlag.

Fausto-Sterling, A. (2003). The problem with sex/gender and nature/nurture, pp. 123–132 in S.J. Williams, L. Birke, and G.A. Bendelow (eds) *Debating biology: Sociological reflections on health, medicine and society*. New York: Routledge.

Feuerstein, G. and R. Kollek. (2000). Risikofaktor Prädiktion. Unsicherheitsdimensionen diagnostischer Humanexperimente am Beispiel prädiktiver Brustkrebstests, pp. 91–115 in L. Honnefelder and C. Streffer (eds) *Jahrbuch für Wissenschaft und Ethik* 5. Berlin: De Gruyter.

Finkler, K. (2000). *Experiencing the new genetics: Family and kinship on the medical frontier.* Philadelphia: University of Philadelphia Press.

Finkler, K., C. Skrzynia, and J.P. Evans. (2003). The new genetics and its consequences for family, kinship, medicine and medical genetics. *Social Science and Medicine* 57: 403–412.

Fitzgerald, J. (1998). Geneticizing disability: The Human Genome Project and the commodification of self. *Issues in Law and Medicine* 14(2): 147–163.

Fleck, L. (1981). *Genesis and development of a scientific fact.* Chicago: University of Chicago Press.

Fleming, R.E. and P. Ponka. (2012). Iron overload in human disease. *New England Journal of Medicine* 366: 348–359.

Foucault, M. (1998). Life: Experience and science, pp. 465–478 in P. Rabinow and J.D. Faubion (eds) *Aesthetics, method, and epistemology: Essential works of Michel Foucault II.* New York: The New Press.

Foucault, M. (2000). Questions of method, pp. 223–238 in P. Rabinow and J.D. Faubion (eds) *Power: Essential works of Michel Foucault III.* New York: The New Press.

Foucault, M. (2003). *The birth of the clinic: An archaeology of medical perception.* New York: Routledge.

Foucault, M. (2007). *Security, territory, population: Lectures at the Collège de France, 1977–78.* New York: Palgrave.

Foucault, M. (2008). *The birth of biopolitics: Lectures at the Collège de France, 1978–79.* New York: Palgrave.

Frankfurter Rundschau. (2004a). Erblich belastete Lehrerin wird Beamtin. *Frankfurter Rundschau* August 3, 2004.

Frankfurter Rundschau. (2004b). Zu dick für Beamtenjob. *Frankfurter Rundschau* June 3, 2004.

Freidson, E. (1979). *Der Ärztestand: Berufs- und wissenschaftssoziologische Durchleuchtung einer Profession.* Stuttgart: Ferdinand Enke Verlag.

Fujimura, J.H. (1996). *Crafting science: A sociohistory of the quest for the genetics of cancer.* Cambridge, MA: Harvard University Press.

Funk, V. (2009). DNA-test für Flüchtlinge. *Frankfurter Rundschau* March 9, 2009: 1.

Galvin, R. (2002). Disturbing notions of chronic illness and individual responsibility: Towards a genealogy of morals. *Health* 6(2): 107–137.

Ganten, D. and K. Ruckpaul. (2001). Molekulare Medizin, pp. 3–19 in A.M. Raem, R.W. Braun, and H. Fenger (eds) *Gen-Medizin. Eine Bestandsaufnahme.* Berlin: Springer.

Gaserow, V. (2007). Grenzen dicht für Birmaner. *Frankfurter Rundschau*, October 22, 2007: 8.

Gaudillière, J.-P. (1995). Sequenzieren, Zählen und Vorhersehen. Praktiken einer Genverwaltung. *Tüte*: 34–39.

Gaudillière, J.-P. (2000). Mendelism and medicine: Controlling human inheritance in local contexts, 1920–1960. *Comptes Rendus de l'Academie des Sciences Serie III–Sciences de la Vie–Life Sciences* 323(12): 1117–1126.

Gaudillière, J.-P. and I. Löwy (eds). (2001). *Heredity and infection: The history of disease transmission.* London: Routledge.

Gavaghan, C. (2007). *Defending the genetic supermarket: Law and ethics of selecting the next generation.* London: Routledge.

Geelen, E., K. Horstman, C. Marcelis, P.A. Doevendans, and I. Van Hoyweghen. (2012). Unravelling fears of genetic discrimination: An exploratory study of Dutch HCM families in an era of genetic non-discrimination acts. *European Journal of Human Genetics.* Advance online publication March 28, 2012. http://www.nature.com/ejhg/journal/vaop/ncurrent/full/ejhg201253a.html

Geetter, J.S. (2002). Coding for change: The power of the human genome to transform the American health insurance system. *American Journal of Law & Medicine* 28: 1–76.

Gelinsky, K. (2008). Warum Forscher Lobbyarbeit für das Recht auf Nichtwissen leisten: Das amerikanische Gesetz gegen genetische Diskriminierung. *Frankfurter Allgemeine Zeitung*, June 2, 2008: 40.

Geller, L.N. (2002). Current developments in genetic discrimination, pp. 267–285 in J.S. Alper, C. Ard, A. Ash, P. Conrad, L.N. Geller, and J. Beckwith (eds) *The double-edged helix: Social implications of genetics in a diverse society*. Baltimore: John Hopkins University Press.

Geller, L.N., J.S. Alper, P.R. Billings, C.I. Barash, J. Beckwith, and M.R. Natowicz (1996). Individual, family, and societal dimensions of genetic discrimination: A case study analysis. *Science and Engineering Ethics* 2: 71–88.

GeneWatch UK. (2001). *Genetic testing in insurance and employment: A new form of discrimination* (Briefing Number 15). Available online at: http://www.genewatch.org/uploads/f03c6d66a9b354535738483c1c3d49e4/brief15.pdf, (accessed December 13, 2012).

GeneWatch UK. (2003). *Genetic testing in the workplace: Creating a genetic underclass?* (Briefing Number 24). Available online at: http://www.genewatch.org/uploads/f03c6d66a9b354535738483c1c3d49e4/Brief24.pdf, (accessed December 13, 2012)

Georgescu, V. and M. Vollborn. (2004). Massen-Gentest. Krankenkasse untersucht 6000 Mitglieder auf Erbkrankheit. *Spiegel Online* August 23, 2004. http://www.spiegel.de/wissenschaft/mensch/massen-gentest-krankenkasse-untersucht-6000-mitglieder-auf-erbkrankheit-a-314596.html (accessed August 24, 2012).

Geppert, C. M.A. and L.W. Roberts. (2005). Ethical issues in the use of genetic information in the workplace: A review of recent developments. *Current Opinion in Psychiatry* 18(5): 518–524.

Gerards, J.H., A.-W. Heringa, and H.L. Janssen. (2005). *Genetic discrimination and genetic privacy in a comparative perspective*. Antwerpen: Intersentia.

Gibbon, S. (2002). Re-examining geneticization: Family trees in breast cancer genetics. *Science as Culture* 11: 429–457.

Gigerenzer, G. (2002). *Reckoning with risk: Learning to live with uncertainty*. London: Penguin.

Gin, B.R. (1997). Genetic discrimination: Huntington's disease and the Americans With Disabilities Act. *Columbia Law Review* 97(5): 1406–1434.

Goffman, E. (1986). *Stigma: Notes on the management of spoiled identity*. New York: Simon & Schuster.

Gomolla, M. and F.-O. Radtke. (2002). *Institutionelle Diskriminierung. Die Herstellung ethnischer Differenz in der Schule*. Opladen: Leske und Budrich.

Görlitzer, K.-P. (2004). Gentest-Gesetz nach Wünschen von Kassen, Forschern und Firmen. *Bioskop* 7(28): 8–9.

Gostin, L. (1991). Genetic discrimination: The use of genetically based diagnostic and prognostic tests by employers and insurers. *American Journal of Law & Medicine* 17(1–2): 109–144.

Götte, M. (2004). Gendiagnostik: Je mehr sich aus den Erbanlagen herauslesen lässt, desto lauter wird der Ruf nach einer gesetzlichen Regelung. *Süddeutsche Zeitung* November 26, 2004.

Gottweis, H. (1998). *Governing molecules: The discursive politics of genetic engineering in Europe and the United States*. Cambridge, MA: MIT Press.

Greely, H.T. (2007). The uneasy ethical and legal underpinnings of large-scale genomic biobanks. *Annual Review of Genomics and Human Genetics* 8: 343–364.

Greenwald, J. (2010). Discrimination claims rising in wake of genetic bias law. *Business Insurance* 44(20): 3–5.

Griffiths, P.E. (2001). Developmental systems theory, pp. 1–4 in *Encylopedia of life sciences*. London: Nature Publishing Group.

Gudding, G. (1996). The phenotype/genotype distinction and the disappearance of the body. *Journal of the History of Ideas* 57: 525–545.

Guedes, C. and D. Diniz. (2007). Um caso de discriminação genética: o traço falciforme no Brasil. *Physis: Rev. Saúde Coletiva* 17(3): 501–520.

Habermas, J. (2003). *The future of human nature*. Cambridge: Polity Press.

Hacking, I. (1986). Making up people, pp. 222–236 in T. Heller, M. Sosna and D.E. Wellberry (eds) *Reconstructing individualism: Autonomy, individuality, and the self in Western thought*. Stanford, CA: Stanford University Press.

Hall, M., J.E. McEwen, J.C. Barton, A.P. Walker, E.G. Howe, J.A. Reiss, T.E. Power, S.D. Ellis, D.C. Tucker, B.W. Harrison, G.D. McLaren, A. Ruggiero, and E.J. Thomson. (2005). Concerns in primary care population about genetic discrimination by insurers. *Genetics in Medicine* 7(5): 311–317.

Hall, M.A. and S.S. Rich. (2000). Laws restricting health insurers' use of genetic information: Impact on genetic discrimination. *American Journal of Human Genetics* 66: 293–307.

Hallowell, N. (1999). Doing the right thing: Genetic risk and responsibility, pp. 97–120 in P. Conrad and J. Gabe (eds) *Sociological perspectives on the new genetics*. Oxford: Blackwell.

Hallowell, N. (2000). Reconstructing the body or reconstructing the woman? Problems of prophylactic mastectomy for hereditary breast cancer risk, pp. 153–180 in L.K. Potts (ed) *Ideologies of breast cancer: Feminist perspectives*. New York: St. Martin's Press.

Hardin, G. (1974). The moral threat of personal medicine, pp. 85–91 in M. Lipkin Jr. and P.T. Rowley (eds) *Genetic responsibility: On choosing our children's genes*. New York: Plenum Press.

Harris, R. and H.J. Harris. (1995). Primary care for patients at genetic risk. *British Medical Journal* 311: 579–580.

Heath, D., R. Rapp, and K.-S. Taussig. (2004). Genetic citizenship, pp. 152–167 in D. Nugent and J. Vincent (eds) *Companion to the handbook of political anthropology*. Oxford: Blackwell.

Hedgecoe, A. (2001). Schizophrenia and the narrative of enlightened geneticization. *Social Studies of Science* 31(6): 875–911.

Heeney, C., N. Hawkins, J. De Vries, P. Boddington, and J. Kaye. (2010). Assessing the privacy risks of data sharing in genomics. *Public Health Genomics* 14(1): 17–25.

Heggie, V. (2010). Testing sex and gender in sports: Reinventing, reimagining and reconstructing histories. *Endeavour* 34(4): 157–163.

Heinemann, T. and T. Lemke. (2013). Suspect families: DNA kinship testing in German immigration policy. *Sociology* (in press).

Heinemann, T., T. Lemke, and B. Prainsack (eds). (2012). Risky profiles: Societal dimensions of forensic uses of DNA technologies. *Special Issue of New Genetics & Society* 30(3).

Hellman, D. (2003). What makes genetic discrimination exceptional? *American Journal of Law & Medicine* 29: 77–116.

Hellman, D. (2008). *When is discrimination wrong?* Cambridge, MA: Harvard University Press.

Henman, P. (2004). Targeted! Population segmentation, electronic surveillance and governing the unemployed in Australia. *International Sociology* 19(2): 173–191.

Henn, W. (2001). Sind wir alle erbkrank? Zur Normalität des genetisch Abnormen. *Universitas* 56: 266–274.

Henn, W. (2002). Probleme der ärztlichen Schweigepflicht in Familien mit Erbkrankheiten. *Zeitschrift für Medizinische Ethik* 48: 343–354.
Hennen, L., T. Petermann, and A. Sauter. (2001). *Das genetische Orakel. Prognosen und Diagnosen durch Gentests—eine aktuelle Bilanz.* Berlin: Edition Sigma.
Hercher, L. (2010). Gender verification: A term whose time has come and gone. *Journal of Genetic Counseling* 19(6): 551–553.
Hindmarsh, R. and B. Prainsack (eds). (2010). *Genetic suspects: Global governance of DNA profiling and databasing.* Cambridge: Cambridge University Press.
Hock, K.T., K.D. Christensen, B.M. Yashar, S.J. Roberts, S.E. Gollust, and W.R. Uhlmann. (2011). Direct-to-consumer genetic testing: An assessment of genetic counselors' knowledge and beliefs. *Genetics in Medicine* 13(4): 325–331.
Hodge, J.G. (2004). Ethical issues concerning genetic testing and screening in public health. *American Journal of Medical Genetics* 125C(1): 66–70.
Hoedemaekers, R. (1998). *Normative determinants of genetic screening and testing.* Delft: Eburon P&L.
Hoedemaekers, R. and H. Ten Have. (1999). The concept of abnormality in medical genetics. *Theoretical Medicine and Bioethics* 20(6): 537–561.
Hoffmann, W. (1999). Die individuelle Empfänglichkeit—ein neuer Faktor im Arbeitsschutz? Einführung und Überblick, pp. 17–34 in R. Frentzel-Beyme and J. Tholen (eds) *Krebsgefährdung am Arbeitsplatz. Individuelle Unterschiede.* Bremen: Universität Bremen.
Hofmann, C. (1999). *Rechtsfragen der Genomanalyse.* Frankfurt am Main: Peter Lang.
Höhn, H. (1997). Die Genetifizierung der Medizin und die Zukunft der Humangenetik. *Medizinische Genetik* 9: 173–174.
Holtzman, N.A. (1989). *Proceed with caution: Predicting genetic risks in the recombinant DNA era.* Baltimore: John Hopkins University Press.
Holtzman, N.A. and M.A. Rothstein. (1992). Invited editorial: Eugenics and genetic discrimination. *American Journal of Human Genetics* 50: 457–459.
Hormel, U. and A. Scherr. (2004). *Bildung für die Einwanderungsgesellschaft. Perspektiven der Auseinandersetzung mit struktureller, institutioneller und interaktioneller Diskriminierung.* Wiesbaden: VS Verlag für Sozialwissenschaften.
Huang, S. (2000). The practical problems of post-genomic biology. *Nature Biotechnology* 18: 471–472.
Hubbard, R. and E. Wald. (1997). *Exploding the gene myth: How genetic information is produced and manipulated by scientists, physicians, employers, insurance companies, educators, and law enforcers.* Boston: Beacon Press.
Hubbard, R. and R.C. Lewontin. (1996). Pitfalls of genetic testing. *New England Journal of Medicine* 334: 1192–1194.
Hudson, K.L. (2007). Prohibiting genetic discrimination. *New England Journal of Medicine* 356(20): 2021–2023.
The Huntington's Disease Collaborative Research Group. (1993). A novel gene containing a trinucleotide repeat that is expanded and unstable on Huntington's Disease chromosomes. *Cell* 72: 971–983.
Husted, J. (1997). Autonomy and a right not to know, pp 55–68 in R. Chadwick, M. Levitt, and D. Shickle (eds) *The right to know and the right not to know.* Aldershot, UK: Avebury.
Ireni-Saban, L. (2010). Regulating genetic information in Germany and Israel: towards a policy network comparative analysis. *Journal of Comparative Policy Analysis* 12(4): 351–372.
Irrgang, B. (2001). Der Krankheitsbegriff der prädiktiven Medizin und die humangenetische Beratung, pp. 651–660 in A.M. Raem, R.W. Braun, and H. Fenger (eds) *Gen-Medizin. Eine Bestandsaufnahme.* Berlin: Springer.

Janich, P. (2001). Der Status des genetischen Wissens, pp. 70–89 in L. Honnefelder and P. Propping (eds) *Was wissen wir, wenn wir das menschliche Genom kennen?* Köln: Dumont.

Jarka, M., B. Brosig and H.-E. Richter. (1996). Psychosoziale Probleme bei Huntingtonscher Chorea. *Psychiatrische Praxis* 23: 117–125.

Joly, Y., M. Braker, and M. Le Huynh. (2010). Genetic discrimination in private insurance: Global perspectives. *New Genetics and Society* 29(4): 351–368.

Kang, P.B. (2011). Presymptomatic and early symptomatic genetic testing. *Continuum Lifelong Learning Neurology* 17(2): 343–346.

Kaplan, J.C. and C. Junien. (2000). Genomics and medicine: An anticipation. From Boolean Mendelian genetics to multifactorial molecular medicine. *Comptes Rendus De L'Academie Des Sciences, Serie III—Sciences De La Vie—Life Sciences* 323(12): 1167–1174.

Kaprio, J. (2001). Commentary: Role of other genes and environment should not be overlooked in monogenic disease. *British Medical Journal* 322: 1023.

Kaufmann, F.-X. (1992). *Der Ruf nach Verantwortung. Risiko und Ethik in einer unüberschaubaren Welt.* Freiburg im Br.: Herder.

Kavanagh, A.M. and D.H. Broom. (1998). Embodied risk: My body, myself? *Social Science & Medicine* 46(3): 437–444.

Kay, L.E. (1993). *The molecular vision of life: Caltech, the Rockefeller Foundation, and the rise of the new biology.* New York: Oxford University Press.

Kay, L.E. (2000). *Who wrote the book of life? A history of the genetic code.* Stanford, CA: Stanford University Press.

Keays, D. (2000). When is genetic discrimination justified? *Monash Bioethics Review* 19: 79–88.

Keller, E.F. (1992). Nature, nurture, and the human genome project, pp. 281–299 in D.J. Kevles and L.E. Hood (eds) *The code of codes: Scientific and social issues in the Human Genome Project.* Cambridge, MA: Harvard University Press.

Keller, E.F. (2000). *The century of the gene.* Cambridge, MA: Harvard University Press.

Keller, E.F. (2002). *Making sense of life: Explaining biological development with models, metaphors, and machines.* Cambridge, MA: Harvard University Press.

Kelly, S.E. (2005). "New" genetics meets the old underclass: Findings from a study of genetic outreach services in rural Kentucky, pp. 137–151 in R. Bunton and A. Petersen (eds) *Genetic governance: Health, risk and ethics in the biotech era.* London: Routledge.

Kenen, R.H. (1994). The Human Genome Project: Creator of the potentially sick, potentially vulnerable and potentially stigmatized? pp. 49–64 in I. Robinson (ed) *Social consequences of life and death under high technology medicine.* Manchester: University of Manchester.

Kenen, R.H. (1996). The at-risk health status and technology: A diagnostic invitation and the "gift" of knowing. *Social Science & Medicine* 42: 1545–1553.

Kenen, R.H. and R.M. Schmidt. (1978). Stigmatization of carrier status: social implications of heterozygote genetic screening programs. *American Journal of Public Health* 68: 1116–1120.

Kerr, A. (2000). (Re)constructing genetic disease: The clinical continuum between cystic fibrosis and male infertility. *Social Studies of Science* 30: 847–894.

Kerr, A. and T. Shakespeare. (2002). *Genetic politics: From eugenics to genome.* Oxford: New Clarion Press.

Kevles, D.J. (1995). *In the name of eugenics: Genetics and the uses of human heredity.* Cambridge, MA: Harvard University Press.

Khoury, M.J., W. Burke, and E. Thompson (eds). (2000). *Genetics and public health in the 21st century: Using genetic information to improve health and prevent disease.* Oxford: Oxford University Press.

Kim, Y.-K. (ed). (2009). *Handbook of behavior genetics*. Luxemburg: Springer.

Kitcher, P. (1996). *The lives to come: The genetic revolution and human possibilities*. New York: Simon & Schuster.

Klees, B. (1990). *Der Griff in die Erbanlagen. Verdrängte Probleme der Genomanalyse*. Braunschweig: Steinweg Verlag.

Klees, B. (1992). Schöne neue Welt—oder neuer Sozialdarwinismus? Auswirkungen von Bio- und Gentechnologien auf die Arbeitswelt, pp. 199–250 in A.-D. Stein (ed) *Lebensqualität statt Qualitätskontrolle menschlichen Lebens*. Berlin: Marhold.

Klitzman, R. (2010). Views of discrimination among individuals confronting genetic disease. *Journal of Genetic Counseling* 19(1): 68–83.

Knoppers, B. M., B. Godard, and Y. Joly. (2004). A comparative international overview, pp. 173–195 in M.A. Rothstein (ed) *Genetics and life insurance: Medical underwriting and social policy*. Cambridge, MA: MIT Press.

Knowles, J.H. (1977). Responsibility of individual. *Daedalus* 106(1): 57–80.

Koch, L. (1993). The genefication of medicine and the concept of disease in Hamburger Institut für Sozialforschung (ed) *Diskussionspapier* 1(93). Hamburg: Hamburger Institut für Sozialforschung.

Koch, L. (1999). Predictive genetic medicine—a new concept of disease, pp. 185–195 in E. Hildt and S. Graumann (eds) *Genetics in human reproduction*. Aldershot: Ashgate.

Koch, L. (2000). *Tvangssterilisation i Danmark 1929–67*. Kopenhagen: Gyldendal.

Koch, L. (2002). The government of genetic knowledge, pp. 92–103 in S. Lundin and L. Akesson (eds) *Gene technology and economy*. Lund: Nordic Academic.

Koch, L. (2004). The meaning of eugenics: Reflections on the government of genetic knowledge in the past and the present. *Science in Context* 17(3): 315–331

Koch, L. and D. Semerding. (1994). The sociology of entrenchment: A cystic fibrosis test for everyone? *Social Studies of Medicine* 39(9): 1211–1220.

Kollek, R. (2002). Fragile Kodierung—Genetik und Körperverständnis, pp. 109–120 in E. Kuhlmann and R. Kollek (eds) *Konfigurationen des Menschen. Biowissenschaften als Arena der Geschlechterpolitik*. Opladen: Leske und Budrich.

Kollek, R. and T. Lemke. (2008). *Der medizinische Blick in die Zukunft. Gesellschaftliche Implikationen prädiktiver Gentests*. Frankfurt am Main: Campus.

Konrad, M. (2003). From secrets of life to the life of secrets: Tracing genetic knowledge as genealogical ethics in biomedical Britain. *Journal of the Royal Anthropological Institute* 9: 339–358.

Korobkin, J.D. and R. Rajkumar. (2008). The genetic information nondiscrimination act—a half-step toward risk sharing. *New England Journal of Medicine* 359(4): 335–337.

Kremer B., P. Goldberg, S.E. Andrew, J. Theilmann, H. Telenius, J. Zeisler, F. Squitieri, B. Lin, A. Bassett, E. Almqvist, T.D. Bird, and M.R. Hayden. (1994). A worldwide study of the Huntington's disease mutation: The sensitivity and specificity of measuring CAG repeats. *New England Journal of Medicine* 330(20): 1401–1406.

Kress, N. (1993). *Beggars in Spain*. New York: William Morrow & Co.

Krimsky, S. and Peter S. (2005). *Rights and liberties in the biotech age: Why we need a genetic bill of rights*. Lanham, MD: Rowman and Littlefield.

Kröger, S. (2010). Das neue Gendiagnostikgesetz und seine Auswirkungen auf den rechtlichen Rahmen beim Abschluss von Versicherungsverträgen. *Medizinrecht* 28: 751–756.

Kühl, S. (1994). *The Nazi connection: Eugenics, American racism, and German national socialism*. New York: Oxford University Press.

Kühn, H. (1993). *Healthismus. Eine Analyse der Präventionspolitik und Gesundheitsförderung in den USA*. Berlin: Edition Sigma.

Kühn, H. (1998). "Selbstverantwortung" in der Gesundheitspolitik. *Jahrbuch für kritische Medizin* 30: 7–20.

Kühn, H. (2000). Normative Ätiologie. Zur Herrschaftlichkeit des gesellschaftlichen Krankheitsverständnisses. *Jahrbuch für Kritische Medizin* 34: 11–18.

Kuran, T. and E.J. McCaffery. (2005). Sex differences in acceptability of discrimination. *University of Southern California Legal Studies Research Paper Series.* Available online at: http://papers.ssrn.com/sol3/papers.cfm?abstract_id=771640 (accessed June 13, 2012).

Kutter, S. and A. Ruess. (2001). Genetische Unterklasse. *Wirtschaftswoche* 31: 78–80.

Labisch, A. (2002). Gesundheit und Medizin im molekularen Zeitalter—Eine historische Vorausschau, pp. 128–136 in L. Honnefelder and P. Propping (eds) *Was wissen wir, wenn wir das menschliche Genom kennen?* Köln: Dumont.

Lander, J. (2011). "There is no legislation for human genes": Governance of genetics and insurance in Germany, pp. 127–152 in K. Horstman, E. Dow and B. Penders (eds) *Governance of health care innovation: Excursions into politics, science and citizenship.* Raleigh, NC: Lulu Academic.

Langbehn, D.R., R.R Brinkman, D. Falush, J.S. Paulsen, M.R. Hayden, and International Huntington's Disease Collaborative Group. (2004). A new model for prediction of the age of onset and penetrance for Huntington's disease based on CAG length. *Clinical Genetics* 65: 267–277

Lange, H.W. (2002). Morbus Huntington—Klinik, Diagnose und Therapie. *Psycho* 28: 479–486.

Lapham, V.E., C. Kozma, and J.O. Weiss. (1996). Genetic discrimination: Perspectives of consumers. *Science* 274: 621–624.

Latour, B. (1988). *The pasteurization of France.* Cambridge: Cambridge University Press.

Lebel, R.R. (1977). Approach to notion of genetic responsibility. *American Journal of Human Genetics* 29(6): A67.

Lehming, M. (2001). US-Unternehmen wegen Gentests verklagt. *Der Tagesspiegel* February 12, 2001.

Leicht, R. (2003). Das Prinzip Zahnbürste. Alle reden von Eigenverantwortung. Aber was ist das? Eine kleine Begriffsgeschichte. *Die Zeit* December 22, 2003: 8.

Lemke, T. (1997). *Eine Kritik der politischen Vernunft. Foucaults Analyse der modernen Gouvernementalität.* Hamburg: Argument.

Lemke, T. (2000). Die Regierung der Risiken. Von der Eugenik zur genetischen Gouvernementalität, pp. 227–264 in U. Bröckling, S. Krasmann, and T. Lemke (eds) *Gouvernementalität der Gegenwart. Studien zur Ökonomisierung des Sozialen.* Frankfurt am Main: Suhrkamp.

Lemke, T. (2002). Mutationen des Gendiskurses. Der genetische Determinismus nach dem Humangenomprojekt. *Leviathan* 30(3): 400–425.

Lemke, T. (2003). Die Logik der Lücke. Einige Anmerkungen zur Kritik des genetischen Reduktionismus, pp. 22–31 in M. Pieper and E.G. Rodriguez (eds) *Gouvernementalität. Ein sozialwissenschaftliches Konzept in Anschluss an Foucault.* Frankfurt am Main: Campus.

Lemke, T. (2004a). Die Gene der Frau. Humangenetik als Arena der Geschlechterpolitik. *Feministische Studien* 22: 22–38.

Lemke, T. (2004b). Disposition and determinism—genetic diagnostics in risk society. *The Sociological Review*, 52: 550–566.

Lemke, T. (2004c). *Veranlagung und Verantwortung. Genetische Diagnostik zwischen Selbstbestimmung und Schicksal.* Bielefeld: transcript Verlag.

Lemke, T. (2005). From eugenics to the government of genetic risk, pp. 95–105 in R. Bunton and A. Petersen (eds) *Genetic governance.* New York: Routledge.

Lemke, T. (2006). "Du und deine Gene"—Subjektivierungsprogramme und Verantwortungskonzepte in Gesundheitsratgebern. Pflege und Gesellschaft. *Zeitschrift für Pflegewissenschaft* 11(4): 293–306.

Lemke, T. (2010). Neue Vergemeinschaftungen? Entstehungskontexte, Rezeptionslinien und Entwicklungstendenzen des Begriffs der Biosozialität, pp. 21–41 in K. Liebsch and U. Manz (eds) *Leben mit den Lebenswissenschaften. Wie wird biomedizinisches Wissen in Alltagspraxis übersetzt?* Bielefeld: transcript Verlag.

Lemke, T. (2011). *Foucault, governmentality, and critique*, Boulder, CO: Paradigm.

Lemmens, T. (2000). Selective justice, genetic discrimination, and insurance: Should we single out genes in our laws? *McGill Law Journal* 45(2): 347–412.

Levitt, M. and E. Pieri. (2009). It could just be an additional test couldn't it? Genetic testing for susceptibility to aggression and violence. *New Genetics and Society* 28(2): 189–200.

Lewontin, R. (2001). *The triple helix: Gene, organism and environment*. Cambridge, MA: Harvard University Press.

Lindee, S. (2005). *Moments of truth in genetic medicine*. Baltimore: Johns Hopkins University Press.

Link, B.G. and J.C. Phelan. (2001). Conceptualizing stigma. *Annual Review of Sociology* 27: 363–385.

Lippman, A. (1991). Prenatal genetic testing and screening: Constructing needs and reinforcing inequities. *American Journal of Law & Medicine* 17: 15–50.

Little, P. (2002). *Genetic destinies*. Oxford: Oxford University Press.

Lloyd, E.A. (1998). Normality and variation: The Human Genome Project and the ideal human type, pp. 552–566 in D.L. Hull and M. Ruse (eds) *The philosophy of biology*. Oxford: Oxford University Press.

Lock, M. (2005). Eclipse of the gene and the return of divination. *Current Anthropology* 46 (Supplement): S47–S60.

Lohr, S. (2005). I.B.M. to put genetic data of workers off limits. *New York Times* October 10, 2005.

Lomax, G. (2002). Chevron v. Echazabal: A sobering decision for environmental health research. *Environmental Health Perspectives* 110(9): A504–A505.

Low, L., S. Kind, and T. Wilkie. (1998). Genetic discrimination in life insurance: Empirical evidence from a cross sectional survey of genetic support groups in the United Kingdom. *British Medical Journal* 317: 1632–1635.

Luhmann, N. (1997). *Die Gesellschaft der Gesellschaft*. Frankfurt am Main: Suhrkamp.

Luján, J.L. and L. Moreno. (1996). Biological lottery and social solidarity: The impacts of genetic diagnosis. *Futures* 28(5): 497–500.

Lupton, D. (1993). Risk as moral danger—the social and political functions of risk discourse in public-health. *International Journal of Health Services* 23(3): 425–435.

Lynch, M., S.A. Cole, R. McNally, and K. Jordan (2008). *Truth machine: The contentious history of DNA fingerprinting*. Chicago: University Of Chicago Press.

Markel, H. (1992). The stigma of disease—implications of genetic screening. *American Journal of Medicine* 93: 209–215.

Markl, H. (1989). Vom Sinn des Wissens. *Die Zeit* September 8, 1989: 62.

Marshall, E. (1999). Beryllium screening raises ethical issues. *Science* 285: 178–179.

Marteau, T.M. and H. Drake. (1995). Attributions for disability: The influence of genetic screening. *Social Science and Medicine* 40(8): 1127–1132.

Marteau, T.M. and N.A. Holtzman. (2000). Will genetics revolutionize medicine? *New England Journal of Medicine* 343(2): 141–144.

Martindale, D. (2001). Pink slip in your genes. *Scientific American* 284(1): 13–14.

Matloff, E. T., H.L. Shappell, K. Brierley, B.A. Bernhardt, W. Mckinnon, and B.N. Peshkin. (2000). What would you do? Specialists' perspectives on cancer genetic testing, prophylactic surgery, and insurance discrimination. *Journal of Clinical Oncology* 18(12): 2484–2492.

Matloff, E.T., K.L. Brierley, H.L. Shappell, B.A. Bernhardt, W. McKinnon, and B.N. Peshkin. (1999). Fear of genetic discrimination among providers of cancer genetic counseling. *American Journal of Human Genetics* 65(4): 60.

Maurer, J. and H. Lehrach. (2001). Geleitwort, pp. xxiii–xxiv in A.M. Raem, R.W. Braun, and H. Fenger (eds) *Gen-Medizin. Eine Bestandsaufnahme*. Berlin: Springer.

Mayor, S. (2003). Report warns of danger of genetic discrimination in the workplace. *British Medical Journal* 327(7417): 702c.

McBride, C.M., D. Bowen, L.C. Brody, C.M. Condit, R.T. Croyle, M. Gwinn, M.J. Khoury, L.M. Koehly, B.R. Korf, T.M. Marteau, K. McLeroy, K. Patrick, and T.W. Valente. (2010). Future health applications of genomics: Priorities for communication, behavioral, and social sciences research. *American Journal of Preventive Medicine* 38(5): 556–565.

M'Charek, A. (2005). *The Human Genome Diversity Project: An ethnography of scientific practice*. Cambridge: Cambridge University Press.

M'Charek, A., V. Toom, and B. Prainsack. (2011). Bracketing off populations does not advance ethical reflection on EVCs: A reply to Kayser and Schneider. *Forensic Science International: Genetics*. doi:10.1016/j.fsigen.2010.12.012.

McGleenan, T. and U. Wiesing. (2000). Insurance and genetics: European policy options. *European Journal of Health Law* 7: 367–385.

McKie, R. (2004). Inventor warns over abuse of DNA data. *Observer* August 8, 2004.

Mechan-Schmidt, F. (2003). Teacher protests at gene bias. *Times Educational Supplement* November 14. Available online at: http://www.tes.co.uk/teaching-resource/Teacher-protests-at-gene-bias-386956/ (accessed August 24, 2012).

Medizinische Genetik. (1994). Internationale Richtlinien zur Durchführung prädiktiver genetischer Diagnostik bei Huntington-Krankheit. *Medizinische Genetik* 6: 405–409.

Meyer, I. (2001). *"Der Mensch als Datenträger?" Zur verfassungsrechtlichen Bewertung postnataler genetischer Untersuchungen*. Berlin: Berlin Verlag.

Mitchell, S. (2005). Public fears genetic discrimination. *Washington Times* May 27, 2005.

Mill, J.S. (1963). *On liberty*. Charleston, SC: Forgotten Books.

Miller, C.T. and B. Major. (2000). Coping with stigma and prejudice, pp. 243–272 in T.F. Heatherton (ed) *The social psychology of stigma*. New York: The Guildford Press.

Miller, P.S. (2005). Analyzing genetic discrimination in the workplace, pp. 173–177 in S. Krimsky and P. Shorett (eds) *Rights and liberties in the biotech age: Why we need a genetic bill of rights*. Lanham, MD: Rowman and Littlefield.

Miller, P. and N. Rose. (2008). *Governing the present*. Cambridge: Polity Press.

Minkler, M. (1999). Personal responsibility for health? A review of the arguments and the evidence at century's end. *Health Education & Behavior* 26: 121–140.

Minkoff, H. and J. Ecker. (2008). Genetic testing and breach of patient confidentiality: Law, ethics and pragmatics. *American Journal of Obstetrics & Gynecology* 198(5): 498.e1–498.e4.

Miringoff, M.-L. (1991). *The social costs of genetic welfare*. New Brunswick, NJ: Rutgers University Press.

Morris, D.B. (1998). *Illness and culture in the postmodern age*. Berkeley, CA: University of California Press.

Morris, L.A. (2010). Life insurance and genetic tests: Risks for insurers and society. *Health, Risk & Society* 12(3): 251–270.

Morrison, P.J. (2005). Insurance, unfair discrimination, and genetic testing. *The Lancet* 9489: 877–879.

Moss, L. (2003). *What genes can't do*. Cambridge, MA: MIT Press.

Muller, M. (2006). Geschlecht als Leistungsklasse. Der kleine Unterschied und seine großen Folgen am Beispiel der "gender verifications" im Leistungssport. *Zeitschrift für Soziologie* 35(5): 392–412.

Müller-Hill, B. (1988). *Murderous science: Elimination by scientific selection of Jews, Gypsies, and others, Germany 1933–1945*. Oxford: Oxford University Press.

Müller-Wille, S. and H.-J. Rheinberger. (2009). *Das Gen im Zeitalter der Postgenomik. Eine wissenschaftshistorische Bestandsaufnahme*. Frankfurt am Main: Suhrkamp.

Murray, T.H. (1997). Genetic exceptionalism and "future diaries": Is genetic information different from other medical information?, pp. 60–73 in M.A. Rothstein (ed) *Genetic secrets: Protecting privacy and confidentiality in the genetic era*. New Haven, CT: Yale University Press.

My DNA Media. (2004). myDNA media endorses passage of the Genetic Non-Discrimination Act of 2003. Available online at: http://www.thefreelibrary.com/myDNA+Media+Endorses+Passage+of+the+Genetic+Nondiscrimination+Act+of. . .-a0118422329 (accessed December 13, 2012).

National Council on Disability. (2003). *Policy brief series: Righting the ADA 9*. Available online at: https://docs.google.com/viewer?a=v&q=cache:wGw4TRQ_5jgJ:www.ncd.gov/rawmedia_repository/06fc86d7_7c3a_4558_8b33_6dee30ab6360?document.pdf+Policy+brief+series:+Righting+the+ADA+9&hl=de&gl=de&pid=bl&srcid=ADGEESgqaYaU46mZ-rCeyMzUZAPXRPMG8WUdYK95Fh-ilBxuoGTce75uSDIMlT6LKoA_sUE9whRrV2D3bRgM65pnaJ5QYtFwJliNlemddkV0WDFO__sFF3-uZy9vUne0SeWJll6qXsZi&sig=AHIEtbQAPXzrTpPcFDyrYcTKz8-LdwOO4g (accessed August 24, 2012).

Nationaler Ethikrat (ed). (2005). *Prädiktive Gesundheitsinformationen bei Einstellungsuntersuchungen*. Berlin: Nationaler Ethikrat.

Nationaler Ethikrat (ed). (2007). *Prädiktive Gesundheitsinformationen beim Abschluss von Versicherungen*. Berlin: Nationaler Ethikrat.

Natowicz, M.R., J.K. Alper, and J.S. Alper. (1992). Genetic discrimination and the law. *American Journal of Human Genetics* 50: 465–475.

Nature Genetics. (2004). "Race" and the human genome. *Nature Genetics Supplement* 36(11).

Nelkin, D. (1992). The social power of genetic information, pp. 177–190, in D.J. Kevles and L. Hood (eds) *The code of codes: Scientific and social issues in the Human Genome Project*. Cambridge, MA: Harvard University Press.

Nelkin, D. and L. Tancredi. (1994). *Dangerous diagnostics: The social power of biological information* (2nd ed). Chicago: University of Chicago Press.

Nelkin, D. and S. Lindee. (1995). *The DNA mystique: The gene as a cultural icon*. New York: W. H. Freemann & Co.

Neuer-Miebach, T. (2001). Genetische Diskriminierung, pp. 53–67 in CDU-Bundesgeschäftsstelle (ed) *Arbeitsmaterialien Bioethik*. Berlin: Author's Edition.

Neumann-Held, E.M. and C. Rehmann-Sutter. (2006). *Genes in development: Re-reading the molecular paradigm*. Durham, NC: Duke University Press.

Newson, A. (2006). Why genetics services should contact at-risk relatives directly. *BioNews*. Available online at: http://www.bionews.org.uk/commentary.lasso?storyid=3185 (accessed August 22, 2012).

New York Times. (2000). *Text of the White House statements on the Human Genome Project* June 27, 2000. Available online at: http://partners.nytimes.com/library/national/science/062700sci-genome-text.html (accessed May 24, 2012).

Niehoff, J.-U. (2002). Leitbilder der Prävention seit den 1970er Jahren, pp. 218–230 in S. Stöckel and U. Walter (eds) *Prävention im 20. Jahrhundert. Historische Grundlagen und aktuelle Entwicklungen in Deutschland*. Weinheim: Juventa.

Novas, C. and N. Rose. (2000). Genetic risk and the birth of the somatic individual. *Economy and Society* 29(4): 485–513.

Nowlan, W.J. (2002). A rational view of insurance and genetic discrimination. *Science* 297: 195–196.

Nowlan, W.J. (2003). A scarlet letter or a red herring? Genetic discrimination is of little concern compared with existing US healthcare problems. *Nature* 421(6921): 313.

Nuffield Council on Bioethics. (1993). *Genetic screening: Ethical issues*. London: Nuffield Council on Bioethics.

Nuffield Council on Bioethics. (2002). *Report on genetics and human behaviour: The ethical context*. London: Nuffield Council on Bioethics. Available online at: http://www.nuffieldbioethics.org/genetics-and-behaviour (accessed August 24, 2012).

O'Malley, P. (1996). Risk and responsibility, pp. 189–207 in A. Barry, T. Osborne, and N. Rose (eds) *Foucault and political reason: Liberalism, neo-liberalism and rationalities of government*. London: UCL Press.

Orentlicher, D. (1997). Genetic privacy in the patient–physician relationship, pp. 77–91 in M.A. Rothstein (ed) *Genetic secrets: Protecting privacy and confidentiality in the genetic era*. New Haven, CT: Yale University Press.

Otlowski, M., K. Barlow-Stewart, S. Taylor, M. Stranger, and S. Treloar. (2007). Investigating genetic discrimination in the Australian life insurance sector: The use of genetic test results in underwriting, 1999–2003. *Journal of Law and Medicine* 14(3): 367–396.

Otlowski, M., S. Taylor, and K. Barlow-Stewart. (2002). Australian empirical study into genetic discrimination. *Eubios: Journal of Asian and International Bioethics*12: 164–167.

Otlowski, M., S. Taylor, and Y. Bombard. (2012). Genetic discrimination: International perspectives. *Annual Review of Genomics and Human Genetics*. Advance online publication May 15, 2012.

Oyama, S. (2000). *The ontology of information*. Durham, NC: Duke University Press.

Oyama, S., P.E. Griffiths and R.D. Gray. (2001). *Cycles of Contingency: Developmental Systems and Evolution*. Cambridge, MA: MIT Press.

Pagnatarro, M.A. (2001). Genetic discrimination and the workplace: Employee's right to privacy v. employer's need to know. *American Business Law Journal* 39(1): 139–185.

Parker, M. and A. Lucassen. (2002). Working towards ethical management of genetic testing. *The Lancet* 360: 1685–1687.

Paslack, R. and J. Simon. (2005). Reaktionen des Rechts auf genetische Diskriminierung und ihre ethische Begründung, pp. 123–152 in W. van den Daele (ed) *Biopolitik*. Wiesbaden: VS Verlag für Sozialwissenschaften.

Paul, D.B. (1994). Eugenic anxieties, social realities, and political choices, pp. 142–154 in C.F. Cranor (ed) *Are genes us? The social consequences of the new genetics*. New Brunswick, NJ: Rutgers University Press.

Paul, D.B. (1998a). A debate that refuses to die, pp. 81–93 in D. Paul, *The politics of heredity: Essays on eugenics, biomedicine, and the nature–nurture debate*. Albany: State University of New York Press.

Paul, D.B. (1998b). Genes and contagious disease: The rise and fall of a metaphor, pp. 157–172 in D. Paul, *The politics of heredity*. Albany: State University of New York Press.

Paul, D.B. (1998c). PKU screening: Competing agendas, converging stories, pp. 173–186 in D. Paul, *The politics of heredity: Essays on eugenics, biomedicine, and the nature–nurture debate*. Albany: State University of New York.

Paul, D.B. (1999). What is a genetic test, and why does it matter? *Endeavour* 23: 159–161.

Pergament, E. (1997). A Clinical Geneticist Perspective of the Patient-Physician Relationship, pp. 92–107 in M.A. Rothstein (ed) *Genetic Secrets: Protecting Privacy and Confidentiality in the Genetic Era*. New Haven, CT: Yale University Press.

Peters, K., K. Apse, A. Blackford, B. McHugh, D. Michalic, and B. Biesecker. (2005). Living with Marfan syndrome: Coping with stigma. *Clinical Genetics* 69(1): 6–14.

Petersen, A. and R. Bunton. (2002). *The new genetics and the public's health*. London: Routledge.

Petrila, J. (2001). Genetic risk: The new frontier for the duty to warn. *Behavioral Sciences & the Law* 19(3): 405–421.

Pettigrew, T.F. and M.C. Taylor. (2000). Discrimination, pp. 688–695 in E.F. Borgatta and R.J.V. Montgomery (eds) *Encyclopedia of sociology* (2nd ed). New York: MacMillan.

Pieri, E. and M. Levitt. (2008). Risky individuals and the politics of genetic research into aggressiveness and violence. *Bioethics* 22(9): 509–518.

Pletke, M. (1997). *Die Zulässigkeit von Genomanalysen an Arbeitnehmern im deutschen und US-amerikanischen Recht*. Aachen: Shaker Verlag.

Pollitz, K., B.N. Peshkin, E. Bangit, and K. Lucia. (2007). Genetic discrimination in health insurance: Current legal protections and industry practices. *Inquiry* 44(3): 350–368.

Porter, R. (1997). *The greatest benefit to mankind: A medical history of humanity from antiquity to the present*. London: HarperCollins.

Prainsack, B. (2008). What are the stakes? Genetic non-discrimination legislation and personal genomics. *Personalized Medicine* 5(5): 415–418.

Proctor, R.N. (1995). *Cancer wars: How politics shapes what we know and don't know about cancer*. New York: Basis Books.

Raem, A.M., R.W. Braun, and H. Fenger. (2001). *Gen-Medizin. Eine Bestandsaufnahme*. Berlin: Springer.

Ramser, J. and A. Meindl. (2001). Die Bedeutung des Humangenomprojekts für die Identifizierung menschlicher Erkrankungsgene, pp. 43–62 in A.M. Raem, R.W. Braun, and H. Fenger (eds) *Gen-Medizin. Eine Bestandsaufnahme*. Berlin: Springer.

Rapp, R. (2000). *Testing women, testing the fetus: The social impact of amniocentesis in America*. New York: Routledge.

Ravina, B., M. Romer, R. Constantinescu, K. Biglan, A. Brocht, K. Kieburtz, I. Shoulson, and M.P. McDermott. (2008). The relationship between CAG repeat length and clinical progression in Huntington's disease. *Movement Disorders* 23: 1223–1227.

Rees, J. (2002). Complex disease and the new clinical sciences. *Science* 296: 698–701.

Reilly, P.R. (1999). Genetic discrimination, pp. 106–132 in C. Long (ed) *Genetic testing and the use of information*. Washington, DC: AEI Press.

Reiser, S.J. (1985). Responsibility for personal health—a historical perspective. *Journal of Medicine and Philosophy* 10: 7–17.

Rembis, M.A. (2009). (Re)defining disability in the "genetic age": Behavioral genetics, "new" eugenics and the future of impairment. *Disability & Society* 24(5): 585–597.

Rheinberger, H.-J. (1996). Jenseits von Natur und Kultur. Anmerkungen zur Medizin im Zeitalter der Molekularbiologie, pp. 287–279 in: Cornelius Borck (ed) *Anatomien medizinischen Wissens*. Frankfurt am Main: Fischer.

Rheinberger, H.-J. (1997a). *Toward a history of epistemic things: Synthesizing proteins in the test tube*. Stanford, CA: Stanford University Press.

Rheinberger, H.-J. (1997b). Von der Zelle zum Gen. Repräsentationen der Molekularbiologie, pp. 265–279 in H.-J. Rheinberger, M. Hagner, and B. Wahrig-Schmidt (eds) *Räume des Wissens, Repräsentation, Codierung, Spur*. Berlin: Akademie Verlag.

Rheinberger, H.-J. (1998). Bemerkungen zur Geschichte der Molekularbiologie, pp. 58–64 in Kunst- und Ausstellungshalle der Bundesrepublik Deutschland (ed) *Gen-Welten*. Köln: Dumont.

Rheinberger, H.-J. (2000). Beyond nature and culture: Modes of reasoning in the age of molecular biology and medicine, pp. 19–30 in M. Lock, A. Young, and A. Cambrosio (eds) *Living and working with the new medical technologies*. Cambridge: Cambridge University Press.

Rhodes, R. (1998). Genetic links, family ties, and social bonds: Rights and responsibilities in the face of genetic knowledge. *Journal of Medicine and Philosophy* 23(1): 10–30.

Rifkin, J. (1998). *The biotech century: Harnessing the gene and remaking the world*. New York: Jeremy P. Tarcher/Putnam.

Rifkin, J. (2000). Genetische Diskriminierung. Eine neue Form des sozialen Vorurteils. *Süddeutsche Zeitung* June 29, 2000.

Roberts, J.L. (2010). Preempting discrimination: Lessons from the Genetic Information Nondiscrimination Act. *Vanderbilt Law Review* 63(2): 439–490.

Robertson, A. (2001). Biotechnology, political rationality and discourses on health risk. *Health* 5(3): 293–309.

Robinson, P.N., P. Krawitz, and S. Mundlos. (2011). Strategies for exome and genome sequence data analysis in disease-gene discovery projects. *Clinical Genetics* 80: 127–132.

Robotham, J. (2004). Genetic tests taken under pressure, survey finds. *The Sydney Morning Herald* November 5, 2004. Available online at: http://www.smh.com.au/news/National/Genetic-tests-taken-under-pressure-survey-finds/2004/11/04/1099547322551.html?oneclick=true (accessed August 24, 2012).

Roll-Hansen, N. and G. Broberg. (1996). *Eugenics and the welfare state: Sterilization policy in Denmark, Sweden, Norway, and Finland*. East Lansing: Michigan University Press.

Roos, B. (1999). *Die genetische Analyse von Stellenbewerbern und das vorvertragliche Informationsstreben des Arbeitgebers*. Sinzheim: Pro Universitate Verlag.

Roos, R. (2010). Huntington's disease: A clinical review. *Orphanet Journal of Rare Diseases* 5: 40.

Rose, N. (1999). *Powers of freedom: Reframing political thought*. Cambridge: Cambridge University Press.

Rose, N. (2007). *The politics of life itself: Biomedicine, power, and subjectivity in the twenty-first century*. Princeton, NJ: Princeton University Press.

Rose, N. and C. Novas. (2005). Biological citizenship, pp. 439–463 in A. Ong and S.J. Collier (eds) *Global assemblages: Technology, politics, and ethics as anthropological problems*. Oxford: Blackwell.

Rosenblatt A., RR. Brinkman, K.Y. Liang, E.W. Almqvist, R.L. Margolis, C.Y. Huang, M. Sherr, M.L. Franz, M.H. Abbott, M.R. Hayden, and C.A. Ross. (2001). Familial influence on age of onset among siblings with Huntington disease. *American Journal of Medical Genetics* 105(5): 399–403.

Rosenstock, I.M., National Research Council (U.S.) Committee for the Study of Inborn Errors of Metabolism, Simopoulos A.P., and Childs B. (1975). *Genetic screening: A study of the knowledge and attitudes of physicians*. Washington, DC: National Academy of Sciences.

Rothblatt, M. (1997). *Unzipped genes: Taking charge of baby-making in the new millennium*. Philadelphia: Temple University Press.

Rothstein, M.A. (ed). (1997). *Genetic secrets: Protecting privacy and confidentiality in the genetic era*. New Haven, CT: Yale University Press.

Rothstein, M.A. (2001). Predictive genetic testing for Alzheimer's disease in long term care insurance. *Georgia Law Review* 35: 707–733.

Rothstein, M. A. (ed). (2004). *Genetics and life insurance: Medical underwriting and social policy*. Cambridge, MA: MIT Press.

Rothstein, M.A. (2008a). Currents in contemporary ethics: GINA, the ADA, and the genetic discrimination in employment. *Journal of Law, Medicine & Ethics* 36(4): 937–840.

Rothstein, M.A. (2008b). Keeping your genes private. Scientific American 299(3): 64–69.

Rothstein, M.A. and M.R. Anderlik. (2001). What is genetic discrimination, and when and how can it be prevented? *Genetics in Medicine* 3(5): 354–358.

Rothstein, M.A. and Y. Joly. (2009). Genetic information and insurance underwriting: Contemporary issues and approaches in the global economy, pp. 127–144 in P. Atkinson, P. Glasner, and M. Lock (eds) *Handbook of genetics and society*. London: Routledge.

Röttgen, K. (2004). *Der zivilrechtliche Schutz vor Diskriminierung und seine verfahrensrechtliche Gewährleistung*. München: Herbert Utz Verlag.

Rouvroy, A. (2008). *Human genes and neoliberal governance: A Foucauldian critique*. Abingdon, UK: Routledge Cavendish.

Rushton, A.R. (1994). *Genetics and medicine in the United States 1800–1922*. Baltimore: Johns Hopkins University Press.

Sanderson, S.C., S.C. O'Neill, L.A. Bastian, G. Bepler, and C.M. McBride. (2010). What can interest tell us about uptake of genetic testing? Intention and behavior amongst smokers related to patients with lung cancer. *Public Health Genomics* 13: 116–124.

Sass, H.-M. (1994). Der Mensch im Zeitalter von genetischer Diagnostik und Manipulation. Kultur, Wissen und Verantwortung, pp. 339–353 in E.P. Fischer and E. Geißler (eds) *Wieviel Genetik braucht der Mensch? Die alten Träume der Genetiker und ihre heutigen Methoden*. Konstanz: Universitätsverlag Konstanz.

Sass, H.-M. (2003). Patienten- und Bürgeraufklärung über genetische Risikofaktoren, pp. 41–55 in H.-M. Sass und P. Schröder (eds) *Patientenaufklärung bei genetischem Risiko*. Münster: Lit Verlag.

Sayce, L. (1998). Stigma, discrimination and social exclusion: What's in a word? *Journal of Mental Health* 7(4): 331–342.

Scherrer, J. (2012). *Das Gendiagnostikgesetz. Eine Darstellung unter besonderer Berücksichtigung verfassungsrechtlicher Fragen. Recht in den Lebenswissenschaften/Life Sciences and Law 6*. Berlin: Lit Verlag.

Schlich, T. (1996). Die Konstruktion der notwendigen Krankheitsursache: Wie die Medizin Krankheit beherrschen will, pp. 201–229 in C. Borck (ed) *Anatomien medizinischen Wissens. Medizin, Macht, Moleküle*. Frankfurt am Main: Fischer.

Schmedders, M. (2004). *Leben mit der genetischen Diagnose. Psychosoziale Aspekte der Krankheitsprädiktion bei der familiären adenomatösen Polyposis*. Bern: Verlag Hans Huber.

Schmedders, M. (2005). Direktive genetische Beratung durch Ärzte? *Gen-Ethischer Informationsdienst* 21(168): 7–9.

Schmidt, W. (2007). Einreise nur gegen Gen-Nachweis. *die tageszeitung (taz)* October 23, 2007: 7.

Schmidtke, J. (1997). *Vererbung und Ererbtes. Ein humangenetischer Ratgeber*. Reinbek: Rowohlt.

Schmidtke, J. (1998). Die Genetisierung der Medizin. *Public Health Forum* 19: 12.

Schmidtke, J. and K. Sperling. (2003). Genetische Tests auf dem Teststand. *Zeitschrift für Biopolitik* 2: 39–47.

Schmuhl, H.-W. (2008). *The Kaiser Wilhelm Institute for Anthropology, Human Heredity and Eugenics, 1927–1945: Crossing boundaries*. Dordrecht: Springer Science + Business Media B.V.

Schneider, I. (1999). Patente Argumente? *Bioskop* 2(8): 14–15.

Scholz, C. (1995). Biographie und molekulargenetische Diagnostik, pp. 33–72 in E. Beck-Gernsheim (ed) *Welche Gesundheit wollen wir? Dilemmata des medizintechnischen Fortschritts*. Frankfurt am Main: Suhrkamp.

Schwägerl, C. (2004). Ohne meine Beamten. *Frankfurter Allgemeine Zeitung* November 9, 2004.

Schwägerl, C. (2005). Gentests. Prinzip Selbstbestimmung. *Frankfurter Allgemeine Zeitung* January 14, 2005.

Senior, V., T.M. Marteau, and T.J. Peters. (1999). Will genetic testing for predisposition for disease result in fatalism? A qualitative study of parents' responses to neonatal screening for familial hypercholesterolaemia. *Social Science & Medicine* 48: 1857–1860.

Sequeiros, J., E.M. Ramos, J. Cerqueira, M.C. Costa, A. Sousa, J. Pinto-Basto, and I. Alonso. (2010). Large normal and reduced penetrance alleles in Huntington disease: Instability in families and frequency at the laboratory, at the clinic and in the population. *Clinical Genetics* 78(4): 381–387.

Sequeiros, J., M. Paneque, B. Guimaraes, E. Rantanen, P. Javaher, I. Nippert, J. Schmidtke, H. Kääriäainen, U. Kristoffersson, and J.J. Cassiman. (2012). The wide variation of definitions of genetic testing in international recommendations, guidelines and reports. *The Journal of Community Genetics* 3(2): 113–124.

Shakespeare, T. (2003). Rights, risks and responsibilites: New genetics and disabled people, pp. 198–209 in S.J. Williams, L. Birke, and G.A. Bendelow (eds) *Debating biology: Sociological reflections on health, medicine and society*. London: Routledge.

Shaw, M.W. (1987). Presymptomatic testing on Huntington's chorea: A right to know, a right not to know or the duty to know. *American Journal of Medical Genetics* 26: 243–246.

Shell, E.R. (2002). *The hungry gene: The science of fat and the future of thin*. New York: Atlantic Monthly Press.

Shoulson, I. and A.B. Young. (2011). Milestones in Huntington disease. *Movement Disorders* 26(6): 1127–1133.

Siems, D. (2004). Datenschützer: Gentests gesetzlich regeln. *Die Welt* August 25, 2004.

Silvers, A. and M.A. Stein. (2002). An equality paradigm for preventing genetic discrimination. *Vanderbilt Law Review* 55(5): 1341–1395

Simon, J. (2001). *Gendiagnostik und Versicherung. Die internationale Lage im Vergleich*. Baden-Baden: Nomos.

Simpson, J.L., A. Ljungqvist, M.A. Ferguson-Smith, A. de La Chapelle, L.J. Elsas, A.A. Ehrhardt, M. Genel, E.A. Ferris, and A. Carlson. (2000). Gender Verification in the Olympics. *Journal of the American Medical Association* 284(12): 1568–1569.

Slaughter, L. (2008). The Genetic Information Nondiscrimination Act: Why your personal genetics are still vulnerable to discrimination. *Surgical Clinics of North America* 88(4): 723–738.

Smart, L. and D.M. Wegner. (2000). The hidden costs of hidden stigma, pp. 220–242 in T.F. Heatherton, R.E. Kleck, M.R. Hebl, and J.G. Hull (eds) *The social psychology of stigma*. New York: The Guildford Press.

Smith, K.C. (2001). A disease by any other name: Musings on the concept of genetic disease. *Medicine, Health Care and Philosophy* 4: 19–30.

Snelling, P.C. (2012). Saying something interesting about responsibility for health. *Nursing Philosophy* 13: 161–178.

Somek, A. (2003). Genetic discrimination. *Society* 40(6): 35–43.

Sommerville, A. and V. English. (1999). Genetic privacy: Orthodoxy or oxymoron? *Journal of Medical Ethics* 25: 144–150.

Sonnenfeld, C. (2002). *Genanalysen im Betrieb*. Available online at: http://www.labournet.de/diskussion/arbeitsalltag/gh/genanalyse.html (accessed December 13, 2012).

Stein, J.T. (2010). Backdoor eugenics: The troubling implications of certain damages awards in wrongful birth and wrongful life claims. *The Seton Hall Law Review* 40: 1117–1168.

Steinberg, D.L. (1996). Languages of risk: Genetic encryptions of the female body. *Women: A Cultural Review* 7: 259–270.

Steinbock, B. (2009). Wrongful life and procreative decisions. *International Library of Ethics, Law, and the New Medicine* 35(Part IV): 155–178.

Steindor, M. (2001). Risiken und Widersprüche genetischer Screeningprogramme in der GKV. Am Beispiel der Hämochromatose-Screenings der Kaufmännischen Krankenkasse. *Arbeit und Sozialpolitik* 1–2: 33–38.

Steindor, M. (2005). Genetische Screenings auf dem Vormarsch. *Gen-Ethischer Informationsdienst* 21(168): 12–16.

Stephenson, J. (1992). A case of discrimination. *GeneWatch* 7(6): 9.

Stockter, U. (2004). Der unberechenbare Mensch. *Gen-Ethischer Informationsdienst* 20(167): 38–42.

Stotz, K. (2005). Organismen als Entwicklungssysteme, pp. 124–143 in U. Krohs and G. Toepfer (eds) *Philosophie der Biologie. Eine Einführung*, Frankfurt am Main: Suhrkamp.

Strachan, T. and A.P. Read. (1996). *Molekulare Humangenetik*. Heidelberg et al.: Spektrum Akademischer Verlag.

Strohman, R.C. (1997). The coming Kuhnian revolution in biology. *Nature Biotechnology* 15: 194–200.

Stumper, K. (1996). *Informationelle Selbstbestimmung und DNA-Analysen*. Frankfurt am Main: Peter Lang.

Suthers, G.K., J. Armstrong, J. McCormack, and D. Trott. (2005). Letting the family know: Balancing ethics and effectiveness when notifying relatives about genetic testing for a familial disorder. *Journal of Medical Genetics* 43: 665–670.

Tan, M.H.-M. (2009). Advancing civil rights, the next generation: The Genetic Information Nondiscrimination Act of 2008 and Beyond. *Health Matrix* 19: 63–119.

Taupitz, J. (2000). *Genetische Diagnostik und Versicherungsrecht*. Karlsruhe: Verlag Versicherungswirtschaft.

Taylor, S. (1998). A case study of genetic discrimination: Social work and advocacy in a new context. *Australian Social Work* 51: 51–57.

Taylor, S. (2011). A population-based survey in Australia of men's and women's perceptions of genetic risk and predictive genetic testing and implications for primary care. *Public Health Genomics* 14: 325–336.

Taylor, S., M. Otlowski, K. Barlow-Stewart, S. Treloar, M. Stranger, and K. Chenoweth. (2004). Investigating genetic discrimination in Australia: Opportunities and challenges in the early stages. *New Genetics and Society* 23(2): 225–239.

Taylor, S., S. Treloar, K. Barlow-Stewart, M. Stranger, and M. Otlowski. (2008). Investigating genetic discrimination in Australia: A large–scale survey of clinical genetics clients. *Clinical Genetics* 74(1): 20–30.

Temple, L.K.F., R.S. McLeod, S. Gallinger, and J.G. Wright. (2001). Defining disease in the genomics era. *Science* 293: 807–808.

Ten Have, H. (1997). Living with the future: Genetic information and human existence, pp. 87–95 in R. Chadwick, M. Levitt, and D. Shickle (eds) *The right to know and the right not to know*. Aldershot: Avebury.

Ten Have, H. (2004). Geneticization: The sociocultural impact of gentechnology, pp. 80–100 in G. Magill (ed) *Genetics and ethics: An interdisciplinary study*. Saint Louis, MO: Saint Louis University Press.

Tesh, S.N. (1988). *Hidden arguments: Political ideology and disease prevention policy*. New Brunswick, NJ: Rutgers University Press.

Testart, J. (2001). Du bluff génétique à la police moléculaire. *Futuribles* 266: 5–16.

Thauvin-Robinet, C., A. Munck, F. Huet, E. Génin, G. Bellis, E. Gautier, M.-P. Audrézet, C. Férec, G. Lalau, M. Des Georges et al. (2009). The very low penetrance of cystic fibrosis for the R117H mutation: A reappraisal for genetic counselling and newborn screening. *Journal of Medical Genetics* 46(11): 752–758.

Thébaud Mondy, A. (1999). Genetische Diskriminierung am Arbeitsplatz. *Le Monde Diplomatique* May 14, 1999: 7.

Tibben, A., R. Timman, E.C. Bannink, and H.J. Duivenvoorden. (1997). Three-year follow-up after presymptomatic testing for Huntington's disease in tested individuals and partners. *Health Psychology* 16: 20–35.

Tjaden, M. (2001). *Genanalyse als Verfassungsproblem*. Frankfurt am Main: Peter Lang.

Tolmein, O. (2002). Down oder Nicht-Down. *die tageszeitung* August 27, 2000: 15.

Tolmein, O. (2004). Gesundheit auf Probe. *Gen-Ethischer Informationsdienst* 20(165): 36–37.

Traufetter, G. (2003). Geisel der eigenen Gene. *Der Spiegel* 42: 216–218.

Treloar, S., S. Taylor, M. Otlowski, K. Barlow-Stewart, M. Stranger, and K. Chenoweth. (2004). Methodological considerations in the study of genetic discrimination. *Community Genetics* 7: 161–168.

Twiss, S.B. (1974). Ethical issues in genetic screening: Models of genetic responsibility, pp. 225–261 in D. Bergsma (ed) *Ethical, social and legal dimensions of screening for human genetic disease*. New York: Stratton Intercontinental Medical Book Corporation.

Tzortzis, A. (2004). Are you (genetically) up to the job? *Deutsche Welle* October 25, 2004. Available online at: www.dw-world.de/dw/article/0,1564,1369432,00.html (accessed August 24, 2012).

Van den Boer-van den Berg, H.M.A. and A.A. Maat-Kievit. (2001). The whole truth and nothing but the truth, but what is the truth? *Journal of Medical Genetics* 38: 39–42.

Van Dijck, J. (1998). *Imagination: Popular images of genetics*. New York: New York University Press.

Van Hoyweghen, I. (2007). *Risks in the making: Travels in life insurance and genetics*. Amsterdam: Amsterdam University Press.

Van Hoyweghen, I. and K. Horstman. (2008). European practices of genetic information and insurance: Lessons for the Genetic Information Nondiscrimination Act. *Journal of the American Medical Association* 300(3): 326–327.

Van Hoyweghen, I., K. Horstman, and R. Schepers. (2006). Making the normal deviant: The introduction of predictive medicine in life insurance. *Social Science and Medicine* 63: 1225–1235.

Varga, O., S. Soini, H. Kääriäainen, J.J. Cassiman, I. Nippert, W. Rogowski, H. Nys, U. Kristoffersson, J. Schmidtke, and J. Sequeiros. (2012). Definitions of genetic testing in European legal documents. *The Journal of Community Genetics* 3(2): 125–141.

Vogel, F. (1998). Gedanken über die Zukunft der Humangenetik. *Medizinische Genetik* 10: 33–39.

Volz, S. (2003). Diskriminierung von Menschen mit Behinderung im Kontext von Präimplantations- und Pränataldiagnostik, pp. 72–88 in S. Graumann and

K. Grüber (eds) *Medizin, Ethik und Behinderung.* Frankfurt am Main: Mabuse Verlag.

Wailoo, K. (2001). *Dying in the city of the blues: Sickle cell anemia and the politics of race and health.* Chapel Hill: The University of Northern Carolina Press.

Waldschmidt, A. (1996). *Das Subjekt in der Humangenetik. Expertendiskurse zu Programmatik und Konzeption der genetischen Beratung 1945–1990.* Münster: Verlag Westfälisches Dampfboot.

Wardle, J., S. Carnell, C.M. Haworth, and R. Plomin. (2008). Evidence for a strong genetic influence on childhood adiposity despite the force of the obesogenic environment. *The American Journal of Clinical Nutrition* 87(2): 398–404.

Wasserman, D. and R. Wachbroit. (2001). *Genetics and criminal behavior.* Cambridge: Cambridge University Press.

Wehling, P. (2011). Biology, citizenship and the government of biomedicine: Exploring the concept of biological citizenship, pp. 225–246 in U. Bröckling, S. Krasmann, and T. Lemke (eds) *Governmentality: Current issues and future challenges.* New York: Routledge.

Weigel, S. (2002). Das Leben ist unberechenbar. *Die Zeit* July 11, 2002: 31.

Weingart, P. (1992). Politik und Vererbung: Von der Eugenik zur modernen Humangenetik, pp. 28–50 in E. Voland (ed) *Fortpflanzung: Natur und Kultur im Wechselspiel. Versuch eines Dialogs zwischen Biologen und Sozialwissenschaftlern.* Frankfurt am Main: Suhrkamp.

Weingart, P., J. Kroll, and K. Bayertz. (1992). *Rasse, Blut und Gene. Geschichte der Eugenik und Rassenhygiene in Deutschland.* Frankfurt am Main: Suhrkamp.

Wenz, P. (2005). Engineering genetic justice. *Bioethics* 19(1): 1–11.

Weß, L. (1992). Eugenik im Zeitalter der Gentechnologie. Vom Zwang zur freiwilligen Inanspruchnahme, pp. 65–82 in A.-D. Stein (ed) *Lebensqualität statt Qualitätskontrolle menschlichen Lebens.* Berlin: Marhold.

Wexler, N. (1992). Clairvoyance and caution: Repercussions from the Human Genome Project, pp. 211–243 in D.J. Kevles and L.E. Hood (eds) *The code of codes: Scientific and social issues in the Human Genome Project.* Cambridge, MA: Harvard University Press.

White, C. (2006). *Americans' attitudes towards genetic discrimination.* Washington, DC: Office of Biotechnology Activities, National Institute of Health. Available online at: http://oba.od.nih.gov/oba/SACGHS/meetings/March2006/white.pdf (accessed May 24, 2012).

Wiesing, U. (1998). Gene, Krankheit und Moral, pp. 95–99 in Kunst- und Ausstellungshalle der Bundesrepublik Deutschland (ed) *Gen-Welten.* Köln: Dumont.

Wilkinson, R. (2010). Unjustified discrimination: Is the moratorium on the use of genetic test results by insurers a contradiction in terms? *Health Care Analysis* 18(3): 279–293.

Willett, W.C. (2002). Balancing life-style and genomics research for disease prevention. *Science* 296: 695–698.

Williams, J.K., D.L. Schutte, C. Evers, and P.A. Holkup. (2000). Redefinition: Coping with normal results from predictive gene testing for neurodegenerative disorders. *Research in Nursing & Health* 23: 260–269.

Williams, S.J. and N.K. Hayward. (2001). The impact of the Human Genome Project on medical genetics. *Trends in Molecular Medicine* 7(5): 229–231.

Winter, S., H. Fenger, and H.-L. Schreiber. (2001). *Genmedizin und Recht. Rahmenbedingungen und Regelungen für Forschung, Entwicklung. Klinik, Verwaltung.* München: Beck.

Wolbring, G. (2001). *Folgen der Anwendung genetischer Diagnostik für behinderte Menschen.* Gutachten erstellt im Auftrag der Enquete-Kommission des Deutschen Bundestages "Recht und Ethik der modernen Medizin." Available online at http://

webarchiv.bundestag.de/archive/2007/0206/parlament/gremien/kommissionen/archiv14/medi/medi_gut_wol.pdf (accessed December 13, 2012).

Wolbring, G. (2005). Disability rights and genetic discrimination, pp. 178–182 in S. Krimsky and P. Shorett (eds) *Rights and liberties in the biotech age: Why we need a genetic bill of rights*. Lanham, MD: Rowman and Littlefield.

Wolf, N. (2000). Genetische Hoffnungen. Zum Wandel des Krankheitsverständnisses bei Krebs. *Jahrbuch für Kritische Medizin* 34: 61–81.

Wolf, S.M. (1995). Beyond "genetic discrimination": Toward the broader harm of geneticism. *Journal of Law Medicine & Ethics* 23: 345–353.

Wolf, U. (1997). Identical mutations and phenotypic variation. *Human Genetics* 100: 305–321.

Wong, J.G. and F. Lieh-Mak. (2001). Genetic discrimination and mental illness: A case report. *Journal of Medical Ethics* 27(6): 393–397.

WonPat-Borja, A.J., L.H. Yang, and J.C. Phelan. (2012). Eugenics, genetics, and mental illness stigma in Chinese Americans. *Social Psychiatry and Psychiatric Epidemiology* 47(1): 145–156.

Wynne, B. (2005). Reflexing complexity. Post-genomic knowledge and reductionist returns in public science. *Theory, Culture & Society* 22(5): 67–94.

Yoxen, E.J. (1984). Constructing genetic diseases, pp. 41–62 in T. Duster and K. Garrett (eds) *Cultural perspectives on biological knowledge*. Norwood, NJ: Ablex Publishing Corporation.

Zerres, K.(2001). Missbrauch von Gendaten? *Focus* 12: 244–246.

Ziegler, J. and A. Ziegler. (2011). Gendiagnostikgesetz und Versicherung: Anspruch und Wirklichkeit. *Zeitschrift für Versicherungswissenschaft* 100: 29–53.

Zimmern, R.L. (1999). Genetic testing: A conceptual exploration. *Journal of Medical Ethics* 25: 151–156.

Zinke, E. (2003). Der gläserne Mensch in der Arbeitswelt. *Gen-Ethischer Informationsdienst* 19(158): 23–25.

Index